AF352856

Microbiology of Fermented Foods and Beverages

Microbiology of Fermented Foods and Beverages

Editor

Theodoros Varzakas

MDPI • Basel • Beijing • Wuhan • Barcelona • Belgrade • Manchester • Tokyo • Cluj • Tianjin

Editor
Theodoros Varzakas
Food Science and Technology
University of the Peloponnese
Kalamata
Greece

Editorial Office
MDPI
St. Alban-Anlage 66
4052 Basel, Switzerland

This is a reprint of articles from the Special Issue published online in the open access journal *Foods* (ISSN 2304-8158) (available at: www.mdpi.com/journal/foods/special_issues/Fermented_Foods).

For citation purposes, cite each article independently as indicated on the article page online and as indicated below:

LastName, A.A.; LastName, B.B.; LastName, C.C. Article Title. *Journal Name* **Year**, *Volume Number*, Page Range.

ISBN 978-3-0365-1848-0 (Hbk)
ISBN 978-3-0365-1847-3 (PDF)

Contents

About the Editor

Theodoros Varzakas

Theodoros Varzakas is a senior full professor at the Department of Food Science and Technology, University of Peloponnese, Greece, specializing in issues of food technology, food processing/engineering, food quality and safety. He is the section editor in chief of the journal *Foods in Food Security and Sustainability* (2020–present) and the ex-editor in chief for *Current Research in Nutrition and Food Science* (2015–2019). He is also a reviewer and member of the editorial boards of many international journals, has written more than 200 research papers and reviews and has presented more than 160 papers and posters at national and international conferences. He has written and edited six books in Greek and six in English on sweeteners, biosensors, food engineering, and food processing, published by CRC, and participated in many European and national research programs as a coordinator or scientific member.

Preface to "Microbiology of Fermented Foods and Beverages"

Fermented foods are consumed all over the world and their consumption shows an increasing trend. They play many roles, from preservation to food security, improved nutrition and social well-being. Different microorganisms are involved in the fermentation process and the diversity of the microbiome is high.

Fermented foods are food substrates that are invaded or overgrown by edible microorganisms whose enzymes hydrolyze polysaccharides, proteins and lipids to nontoxic products with flavors, aromas, and textures that are pleasant and attractive to the human consumer. Fermentation plays different roles in food processing, including the development of a wide diversity of flavors, aromas, and textures in food, lactic acid, alcoholic, acetic acid, alkaline and high salt fermentations for food preservation purposes, biological enrichment of food substrates with vitamins, protein, essential amino acids, and essential fatty acids and detoxification during food fermentation processing.

Theodoros Varzakas
Editor

Editorial

Microbiology of Fermented Foods and Beverages

Theodoros Varzakas 

Department of Food Science and Technology, University of the Peloponnese, Antikalamos, 24100 Kalamata, Greece; theovarzakas@yahoo.gr or t.varzakas@uop.gr

Received: 30 October 2020; Accepted: 9 November 2020; Published: 13 November 2020

Keywords: fermented foods; microbiology; microbiome; health

Fermented foods are consumed all over the world and show increasing trends. They play many roles, from preservation to food security, improved nutrition and social well-being. Different microorganisms are involved in the fermentation process and the diversity of the microbiome is high.

Varzakas et al. [1] have reported on the different fermented vegetables worldwide and the versatility of microorganisms involved. They highlighted soybean tempe and other soybean paste products, sauerkraut, fermented olives, fermented cucumber and kimchi. Moreover, salting procedures are well explained along with the role of lactic acid bacteria in fermented vegetables.

One of these types of microorganisms involved in fermented foods is lactic acid bacteria, which has a strong antibacterial effect due to the production of bacteriocins [2].

Zabat et al. [3] utilized 16S rRNA amplicon sequencing to profile the microbial community of naturally fermented sauerkraut throughout the fermentation process and analyzed the bacterial communities of the starting ingredients and the production environment. They showed that the sauerkraut microbiome is rapidly established after fermentation begins and that the community is stable through fermentation and packaging for commercial sale.

On the other hand, yeasts such as *Saccharomyces cerevisiae* have been added in the dough of bakery products to improve organoleptic properties and reduce spoilage. In this direction, the potential use of *L. plantarum* UFG 121 in the biomass of the dough has been explored, as a biocontrol agent in bread production and a species- or strain-dependent sensitivity of the molds was suggested to the same microbial-based control strategy [4]. Moreover, Kara Ali [5] studied the production of the biomass of *S. cerevisiae* on an optimized medium using date extract as the only carbon source in order to obtain a good yield of the biomass. The biomass production was carried out according to the central composite experimental design (CCD) as a response surface methodology.

Finally, Bell et al. [6,7], highlighted the role of fermented foods and beverages on gut microbiota and debated for the need of transdisciplinary fields of One Health to enhance communication. They addressed nutritional and health attributes and reported that they are not included globally in world food guidelines. They also referred to some traditional African fermented products.

Fermented foods have well-known uses in human health and could help in the prevention of chronic diseases from the general gut health, to immune support, skin health, cholesterol control and lactose intolerance. More research is required in the direction of consumption of fermented foods, their benefits and daily administration.

Funding: This research received no external funding.

Conflicts of Interest: The authors declare no conflict of interest.

References

1. Varzakas, T.; Zakynthinos, G.; Proestos, C.; Radwanska, M. Fermented Vegetables. In *Minimally Processed and Refrigerated Fruits and Vegetables*; Yildiz, F., Wiley, R.C., Eds.; Springer: Boston, MA, USA, 2017; Chapter 15; pp. 537–584.
2. Agriopoulou, S.; Stamatelopoulou, E.; Sachadyn-Król, M.; Varzakas, T. Lactic Acid Bacteria as Antibacterial Agents to Extend the Shelf Life of Fresh and Minimally Processed Fruits and Vegetables: Quality and Safety Aspects. *Microorganisms* **2020**, *8*, 952. [CrossRef] [PubMed]
3. Zabat, M.A.; Sano, W.H.; Wurster, J.I.; Cabral, D.J.; Belenky, P. Microbial Community Analysis of Sauerkraut Fermentation Reveals a Stable and Rapidly Established Community. *Foods* **2018**, *7*, 77. [CrossRef] [PubMed]
4. Russo, P.; Fares, C.; Longo, A.; Soano, G.; Capozzi, V. *Lactobacillus plantarum* with Broad Antifungal Activity as a Protective Starter Culture for Bread Production. *Foods* **2017**, *6*, 110. [CrossRef] [PubMed]
5. Kara Ali, M.; Outili, N.; Ait Kaki, A.; Cherfia, R.; Benhassine, S.; Benaissa, A.; Kacem Chaouche, N. Optimization of Baker's Yeast Production on Date Extract Using Response Surface Methodology (RSM). *Foods* **2017**, *6*, 64. [CrossRef] [PubMed]
6. Bell, V.; Ferrão, J.; Pimentel, L.; Pintado, M.; Fernandes, T. One Health, Fermented Foods, and Gut Microbiota. *Foods* **2018**, *7*, 195. [CrossRef] [PubMed]
7. Bell, V.; Ferrão, J.; Fernandes, T. Nutritional Guidelines and Fermented Food Frameworks. *Foods* **2017**, *6*, 65. [CrossRef] [PubMed]

 *foods*

Review

One Health, Fermented Foods, and Gut Microbiota

Victoria Bell [1], Jorge Ferrão [2], Lígia Pimentel [3], Manuela Pintado [3] and Tito Fernandes [4,*]

1   Faculdade de Farmácia, Universidade de Coimbra, Azinhaga de Santa Comba, 3000-548 Coimbra, Portugal; victoriabell1103@gmail.com
2   Universidade Pedagógica, Rua João Carlos Raposo Beirão 135, Maputo 1000-001, Mozambique; ljferrao@icloud.com
3   CBQF-Centro de Biotecnologia e Química Fina, Escola Superior de Biotecnologia, Universidade Católica Portuguesa, Rua Arquiteto Lobão Vital, Apartado 2511, 4202-401 Porto, Portugal; lpimentel@porto.ucp.pt (L.P.); mpintado@porto.ucp.pt (M.P.)
4   Faculdade de Medicina Veterinária, Universidade de Lisboa, 1300-477 Lisboa, Portugal
*   Correspondence: profcattitofernandes@gmail.com; Tel.: +351-919927930

Received: 20 October 2018; Accepted: 29 November 2018; Published: 3 December 2018

Abstract: Changes in present-day society such as diets with more sugar, salt, and saturated fat, bad habits and unhealthy lifestyles contribute to the likelihood of the involvement of the microbiota in inflammatory diseases, which contribute to global epidemics of obesity, depression, and mental health concerns. The microbiota is presently one of the hottest areas of scientific and medical research, and exerts a marked influence on the host during homeostasis and disease. Fermented foods and beverages are generally defined as products made by microbial organisms and enzymatic conversions of major and minor food components. Further to the commonly-recognized effects of nutrition on the digestive health (e.g., dysbiosis) and well-being, there is now strong evidence for the impact of fermented foods and beverages (e.g., yoghurt, pickles, bread, kefir, beers, wines, mead), produced or preserved by the action of microorganisms, on general health, namely their significance on the gut microbiota balance and brain functionality. Fermented products require microorganisms, i.e., *Saccharomyces* yeasts and lactic acid bacteria, yielding alcohol and lactic acid. Ingestion of vibrant probiotics, especially those contained in fermented foods, is found to cause significant positive improvements in balancing intestinal permeability and barrier function. Our guts control and deal with every aspect of our health. How we digest our food and even the food sensitivities we have is linked with our mood, behavior, energy, weight, food cravings, hormone balance, immunity, and overall wellness. We highlight some impacts in this domain and debate calls for the convergence of interdisciplinary research fields from the United Nations' initiative. Worldwide human and animal medicine are practiced separately; veterinary science and animal health are generally neither considered nor inserted within national or international Health discussions. The absence of a clear definition and subsequent vision for the future of One Health may act as a barrier to transdisciplinary collaboration. The point of this mini review is to highlight the role of fermented foods and beverages on gut microbiota and debate if the need for confluence of transdisciplinary fields of One Health is feasible and achievable, since they are managed by separate sectors with limited communication.

Keywords: nutrition; probiotics; fermented foods; health benefits

---

1. Introduction

The microbiota exerts a marked influence on the host during homeostasis linked with metabolic diseases in humans, but demonstration of causality remains a challenge [1].

Humans as hosts have co-evolved with microorganisms over millions of years, and each body habitat has a unique set of microorganisms shaping its microbiota [2].

These bacteria live on the skin, in the corners of the eyes, in the oral cavity, under fingernails, and most importantly, in the guts. Several perinatal determinants, such as caesarean section delivery, type of feeding, the use of antibiotics, gestational age or environment can affect the pattern of bacterial colonization, resulting in gut dysbiosis. The establishment and development of the gut microbiota over the lifecycle moved from the previous accepted dogma that the mammalian healthy placenta and foetus were germ-free and considered to be sterile, and that these conditions were critical to the developing newborn's immune system, to the actual knowledge that in utero humans are now known to harbour unique prenatal microbiomes [3,4].

Amniotic fluid may contain microorganisms, increasing the complexity of fetal microbiota, and having implications for the long-term health and susceptibility to disease, as placental microbiota could trigger immune responses in the fetus. Early gut microbiota settlement influences the maturation of the infant's immune system [5] and subsequent health, although the evidence in support of the "in utero colonization hypothesis" is considered extremely weak by some authors [6].

Health authorities are now becoming fully aware that one cannot be considered to be in good health without a well-balanced microbiota composition in the gut, our "forgotten organ" [7], and of the fundamental role of a diverse and healthy gut microbiota on the subsequent maintenance of future health and well-being of the host [8,9]. Indeed, although it is broadly mentioned that there are 10 times more cells from microorganisms in our bodies than there are human cells [10], this claim has been challenged, and others have estimated that the number of bacteria is similar to that of human cells [11].

Many species of bacteria, specifically those found in the invisible universe of the human microbiota, e.g., composed of nonpathogenic commensal microbiota from the *Firmicutes*, *Bacteroidetes*, *Actinobacteria*, *Proteobacteria* and *Verrucomicrobia* phyla [12], are unsusceptible to petri dish cultivation. They can be successfully cultivated in association with other microbes, meaning in communities of different bacteria species. But without being able to isolate them, research is difficult [13]. Commensal microbiota gradually deteriorates in sick patients. Therefore, research is being conducted to generate new technologies to study the rest of the human microbiome using advances in DNA-sequencing technologies and associated computational methods [14]. Metagenomic sequencing of total fecal DNA samples offers complementary support to classical microbiology, and enables researchers to access previously-inaccessible genomic information from gut bacteria [15,16].

In recent years, a number of functional species and strains have been identified in human metabolic diseases [17]. Gut bacteria can produce various bioactive metabolites which can be detrimental to the host's health, such as those with cytotoxicity, genotoxicity, or immunotoxicity [18], shifting the paradigm of understanding the root cause of the onset and progression of several human metabolic diseases [19].

Gut microbiota modulates the expression of many genes in the human intestinal tract [20], including genes involved in immunity, nutrient absorption, energy metabolism, and intestinal barrier function. It is important to understand genomic diversity of specific members of the gut microbiota if precise nutrition-based approaches are to be realized [21].

In the oral use of live bacteria, there is more research concerning isolated probiotic commercial supplements than there is work concerning health benefits of common fermented foods, since major industries usually do not fund this type of research [22]. Many studies suggest that probiotics may help with diarrhea or symptoms of irritable bowel syndrome, but strong evidence to support their use for most health conditions is lacking in people with sepsis, and probiotics are no panacea [23,24].

Probiotics should not be universally given as a 'one-size-fits-all'; most trials were based on stool samples, which may not really reflect the bacteria living in the gut, as shedding takes place continuously [25]. Besides, taking probiotics after treatment with broad spectrum antibiotics may actually delay the return of normal gut microbiome, a new potential adverse side effect [26].

The One Health concept, introduced at the beginning of the 2000s [27], is a worldwide strategy for promoting multidisciplinary partnerships and information in all facets of health care sciences, perceiving the interrelationship between humans, animals, plants, and their common environment [28].

By working with physicians, veterinarians, osteopathic physicians, dentists, pharmacists, nurses, ecologists, wildlife professionals, and other scientific-health and environmentally-related specialists, it will be possible to monitor and control public health threats and learn how diseases spread among people, animals, and the environment [29].

The point of this mini review is to highlight if the requirement for multiconvergence of the research fields of One Health (Human-Animal-Environment), the relationship between microbiota-nutrition and fermented foods, and to underline the idea that future gut-brain research is feasible and achievable (Figure 1).

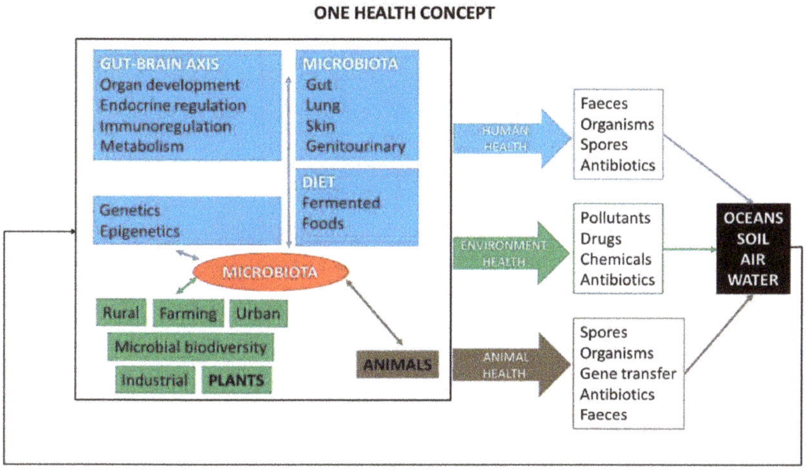

Figure 1. A general picture of the One Health (Human-Animal-Environment) concept as a trans-disciplinary effort. Contribution of the three branches of public health. Microbiota and human metabolic diseases, animal health, and environmental epidemiology.

2. Microbiota and General Health

Having an active and natural variety of microorganisms in the gut may improve general health [30]. The good, healthy bacteria make food more digestible through their enzymes, increased vitamin synthesis, and the preservation of nutrients, and also help to reduce sweet cravings, maintain the immune system, and benefit overall gut wellness [31].

The microbiome, consisting of microorganisms and their collective genomes, modulates the host metabolic phenotype, and influences the host immune system. It is now well established that gut bacteria are closely tied to immune health [32]. The gut microbiota regulates L-tryptophan metabolism and identifies the underlying molecular mechanisms of these interactions [33].

A large majority of the immune system resides in the tonsils and gut, so when gut health is imbalanced, it is hard for the body's immune system to function properly [34]. There are also a number of common factors in modern life that can throw human gut bacteria off, such as processed foods and antibiotics. The use of antibiotics does have several short and long-term implications in the ecology of the normal microbiota and gut motility [35].

Research on the health benefits of probiotics is still emerging, mainly from the food and beverage industries and their commercial interests. In contrast, strong, independent scientific evidence to support specific uses of probiotics for most health conditions is still lacking [36].

The administration of probiotics/prebiotics has been shown to alter the composition and functionality of the gut microbiota [37]. Recent evidence indicates that the effects of probiotics are likely to be different from one person to the next [38].

In addition, probiotics might be ineffective, and possibly counterproductive, in restoring the baseline gut microbiome after it has been altered by antibiotic treatment. Indeed, probiotics may not be quite as good as was commonly thought, and they could even be harmful if taken after antibiotics [39].

Serious disorders such as obesity, anorexia, irritable bowel syndrome, autism, and posttraumatic stress disorder—which have been thought to be solely psychological—share a common symptom: a hypersensitivity to gut stimuli [40–42].

The role of environmental factors in the development of autism is a crucial and an important area of research concerning how the environment influences and interacts with genetic susceptibility. Factors such as parental age at conception, maternal nutrition, infection during pregnancy, and premature birth are risk factors [43]. Autism (ASD, autism spectrum disorder), a developmental disorder characterized by disturbance in language, perception, and socialization, with no exact known cause, is usually linked with bioenergetic metabolism deficiency [44] and neuro-inflammatory conditions [45], and immune system dysregulation and dreadful gut concerns may improve with better diet and fermented foods (e.g., fermented raw coconut milk) [46,47].

Specific benefits from the direct dietary modulation of the human gut microbiota has been described [48]. Despite the wide array of beneficial mechanisms deployed by probiotic bacteria and fermented foods and beverages, relatively few effects have been supported by clinical data [49].

The interactions (Figure 2) between ingested fermented food and intestinal microbiota, and their correlations to metabolomics profiles and health, represent an important perspective, and independent research on health benefits is still emerging [50,51]. Microbiota is specific to each individual, despite the existence of several bacterial species shared by the majority of adults. A diverse and propitious microbial ecosystem (e.g., *Bacteroides fragilis*, *Bifidobacterium* spp. and *Faecalibacterium* spp.) favors homeostasis, particularly at the level of the disease–immune dialogue [52,53].

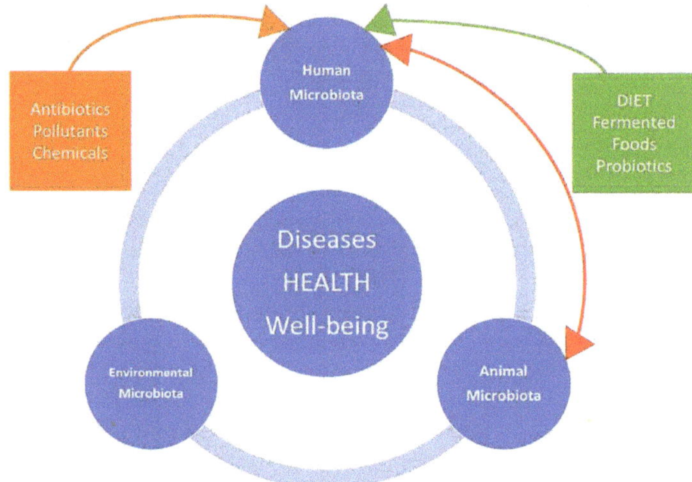

Figure 2. Interactions between dynamics of microbiota in humans, animals, and the associated environment with disease occurrence, salubrity, and well-being.

3. Fermented Foods, Probiotics, Body and Mind

The use of fermentation in conserving food and beverage as a means to provide better taste, improve nutrition and food safety, organically preserve foodstuffs, and promote health properties, is a well-known ancient practice. The reasons for fermenting foods and beverages include improvements of a product's storage time, safety, functionality, organoleptic quality, and nutritional quality properties [54]. Not only is this process beneficial for extending shelf-life, but also, fermentation can enhance nutritional properties in a safe and effective manner [55].

Many types of food groups, including dairy, vegetables, legumes, cereals, starchy roots, and fruits, as well as meat and fish, can be fermented [56]. Fermented foods and beverages can comprise anywhere from 5–40% of the human diet in some populations [57].

Phytochemicals, defined as the non-nutritive, naturally-occurring chemicals found in fruits, vegetables, wholegrains, legumes, beans, herbs, spices, nuts, and seeds, are responsible for producing physiological properties, as well as protecting against various environmental stressors of the plant crops. There are more than one thousand known phytochemicals (e.g., lycopene in tomatoes, isoflavones in soy, and flavanoids in fruits). The microbiota comes into contact with a wide variety of dietary components that escape gut digestion and may be affected by phytochemicals [58].

Substantial confusion exists between fermented foods and beverages and the probiotic concept. It is important to address the common misconception that fermented foods are the same thing as probiotics [59]. They are not probiotics, although they may contain them, as their live microbial content is undefined. The term "probiotic" was first coined [60] in 1974, and many authors have described the history and the progress of probiotics and their different applications. Ilya Ilyich Metchnikoff, the Nobel Prize winner in Medicine in 1908, was the first who observed the effect of what is called now "probiotic" [61]. FAO/WHO redefined the term "probiotics", which is now widely accepted as constituting "live microorganisms that, when administered in adequate amounts, confer a health benefit on the host" [62]. Different types of bacteria (e.g., *Lactobacillus*, *Bifidobacterium*, *Streptococcus*, *Bacillus*) and yeast or mold (e.g., *Saccharomyces*, *Aspergillus*, *Candida*) are used as probiotics. Probably, the first real use of food containing probiotics was fermented milk, but today we have to differentiate between probiotics and probiotic-containing foods (e.g., fermented foods) [63]. The scope and appropriate use of the term "probiotic" has been well clarified (Figure 3) [64].

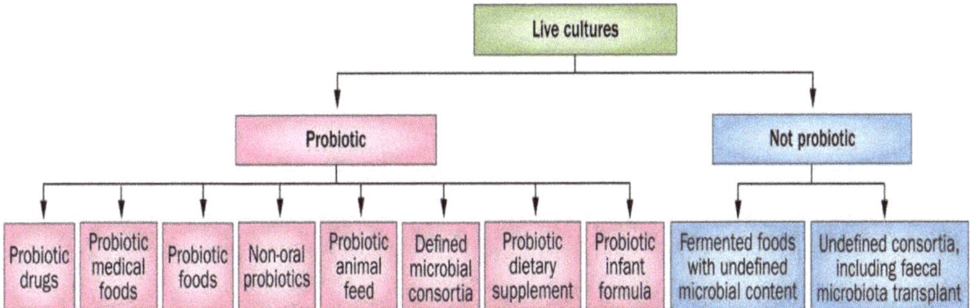

Figure 3. Overall framework for probiotic products [64].

Probiotics are able to renew, restore, and grow affected tissues lining the digestive tract with beneficial microorganisms neutralizing the harmful ones. Useful live microorganisms will regenerate our microflora fermenting our food correctly and improving our health [65,66].

Despite the impact of fermented foods and beverages on gastro-intestinal well-being and diseases, their health benefits or recommended consumption have not been widely translated to global inclusion in world food guidelines [67]. When fermented foods and beverages are supplemented with probiotic bacteria, they provide numerous extra nutritional and health characteristics [68].

Fermented foods and beverages are more popular than ever before, while research into the health benefits of fermented foods is relatively new. Not all fermented foods contain live organisms; beer and wine, for example, undergo steps that remove the organisms, and other fermented foods like bread are heat-treated and the organisms are inactivated. The strain composition and stability of the microbes in fermented foods is not well understood [69].

Fermentation generates adjustments in yeast and live microorganisms cultures in the absence of air, but retains the enzymes, vitamins, and minerals in foods and beverages, which are usually destroyed by processing [70]. The fermenting microorganism, bacteria or yeast, plays a precious role in the functional property of fermented foods and beverages [71]. One the biggest benefits of fermented foods comes from the probiotics they might contain [72]. There are currently no authorized European health claims for probiotics, and the application of probiotics is controversial, since the European Food Safety Authority (EFSA) rejected all submitted health claims related to the term "probiotic", while accepting the term "live microorganism cultures" in yoghurt [73].

Traditional and modern dietary practices utilize fermented foods and beverages, contributing significantly to the food chain value and belonging to a category of foods called "functional foods" (e.g., probiotics, prebiotics, stanols and sterols) by having an additional characteristic, i.e., health-promotion or disease prevention effect [74].

Fermentation converts sugars, in the absence of oxygen, into organic acids, gases, alcohols, and carbon dioxide, and provides several benefits such as new and desirable tastes and textures, enhancement of nutrients (e.g., linoleic acid; bioactive peptides), removal of toxic or undesirable food constituents (e.g., phytic acid; bitter-tasting phenolic compounds), delivery of probiotic bacteria (e.g., *Lactobacillus delbrueckii* subsp. *bulgaricus*; *Streptococcus thermophilus*), and inhibition of foodborne pathogens [75,76].

Fermented foods and beverages are useful because they help provide a spectrum of probiotics to foster a vigorous microbiome. Fermented foods with unidentified microbial content cannot be considered probiotic suppliers. The two main effects of the daily consumption of fermented foods are upon the immune system and upon metabolic function [77].

Dealing with fermented foods has parallels with One Health, since it involves the links between human, animal, environment, foods and microbiota that impacts the organoleptic and physicochemical characteristics of foods as well as human health [78].

There are well documented effects of how adverse early life influences on the gut-brain axis and the use of fermented foods and beverages, mainly with probiotic bacteria, can restore a disturbance of the normal luminal habitat, and so change the effects of the central nervous system on the microbiota [79].

Our guts control and deal with every aspect of our health. How we digest our food, and even the food sensitivities we have are linked to our mood, behavior, energy, weight, food cravings, hormone balance, and immunity [80]. The interaction of nutrients with the microbiota is essentially what determines overall health. Eating and drinking fermented foods and beverages, especially organic unpeeled and unpasteurized fruits and vegetables, improves the bioaccessibility and bioavailability of food bioactive components, supplying dietary fibers and essential micronutrients such as trace-elements and phytochemicals, together with enzymes, lactic acid bacteria, and organic acids, all of which are crucial for good health [81].

Changes in the human colonic microbiota fingerprint are associated with the major causes of morbidity and mortality worldwide, diabetes and cardiovascular diseases, due to imbalances between beneficial and pathogenic bacteria [82].

Physiologically-active peptides with different functionalities are produced from food proteins during fermentation and food digestion by lactic acid bacteria. In some fermented products, bioactive peptides (e.g., immunoglobulins, antibacterial peptides, antimicrobial proteins, oligosaccharides, lipids, and other "minor" components) have the potential to be used in the formulation of health-enhancing nutraceuticals [83], and include short amino acid sequences that, upon release from the parent protein, may play different physiological roles, including antioxidant, antihypertensive, antimicrobial, and other bioactivities [84,85].

Fermentation may enhance the benefits of a wide variety of foods, dairy products, herbs, and beverages, acting upon the absorption and activity of their secondary metabolites and chemical elements [86]. However, it is not always possible to clearly distinguish the potential contribution of the microbial content from that of the food matrix. There is recent evidence and consumer perception of the health benefits of fermented foods and beverages [87], beyond the popular recognized effects on the impairment of gastrointestinal function, namely, their relevance on gut microbiota, correlated to human health and to several infectious [88], inflammatory, and neoplastic disease processes [89], as well as to brain functionality [90].

Despite disagreement among mental health practitioners and researchers pertaining to the aetiology, categorization, and medical care of several mental disorders [91], current research regarding fermented foods, the microbiome, and their effect on human health, particularly the global epidemic of mental health [92], describes problems associated with the modern lifestyle, and with the western diet being high in sugar and saturated fat [93].

The degradation of the intestinal mucous membrane, weakening the tight barrier against the ingress of harmful substances, and the protection against a reaction to omnipresent harmless compounds, is a primary cause of several disturbances [94].

Ingestion of vibrant probiotics, especially in fermented foods, is found to cause significant positive improvements in balancing intestinal permeability and barrier function [95], with direct effects on metabolic syndrome, atherosclerosis, inflammatory bowel diseases, and colon cancer [96] and indirect effects on depression, anger, anxiety, and levels of stress hormones [97].

Young individuals with autism often have a reduced number of microorganisms in the gut [98], and atypical digestive health conditions may occur, like chronic gastrointestinal and functional bowel disorder, causing discomfort, diarrhea and bloating, abdominal pain and cramping, collectively described as irritable bowel syndrome [99]. Children with autism spectrum, besides having a genetic predisposition, show a disruption of the indigenous gut flora and an elevated number of potentially pathogenic (toxin-producing) *Clostridia* in the gut [100,101]. The effectiveness of fermented foods, mushroom biomass, and probiotics in relieving gut symptoms in autistic children has been studied [102–104].

The involvement of the microbiota in inflammatory diseases may contribute to altered mood via intestinal permeability, systemic and local lipopolysaccharide burden, and even direct-to-brain microbial communication [105]. In future, insights based upon omics techniques will increase our knowledge between pathogens and healthy strains, thereby explaining food ecosystems and their dynamics [106,107].

4. Fermented Foods in Developing Settings

Around the world, each culture has its own distinctiveness in terms of food culture and heritage, where fermented foods are included. In the developing world, for people living in poverty, the main priority is not food hygiene, safety, and nutritive factors, as they consume less nutritious foods in which chemical, microbiological, zoonotic, and other hazards may pose a health risk [108].

African traditional fermented foods and beverages have been used since ancient times. Throughout the continent, there is great variety of fermented foods and beverages, mainly sour porridges and drinks. Of the various types of fermentations used to obtain fermented foods and beverages, lactic acid and alcoholic fermentations are the most popular in developing settings, where some 80% of the population still seek care from traditional healers who prescribe indigenous products.

In Africa, a continent which consumes high levels of lactic acid fermented products, estimates for mental disorders and depression vary widely, but seemingly, such diseases are not less common than in developed societies [109], although factors other than diet exist which may exacerbate conditions such as socio-economic changes, urbanicity, alterations in dietary habits, and, more recently, sedentary behavior among youth [110].

People from Sub-Saharan Africa, often plagued by civil conflicts, drought, floods, famine, and disease, but with huge biodiversity of plants and herbs, tend to rely on traditional healers who often interpret mental illness in terms of possession or curses, and tackle mental health by rituals, but also by recommending traditional plants, herbs, fermented foods, and beverages [111,112]. Many rural communities in Africa are totally reliant on traditional fermented foods as the primary source of nutrition for nourishment, as well as for cultural traditional practices [113].

In Mozambique and Zimbabwe, traditional fermented foods are used for weaning from the age of four months. The commonest fermented food is known are *mahewu*, a traditional, fermented, malted, sour, non-alcoholic maize or cassava thin porridge, sour milk and sour porridge [114,115]. The Tanzanian fermented gruel, *togwa*, has been found to protect against foodborne illnesses in regions that have poor sanitation [116].

However, each person is unique in their needs and sensitivities. Most of us only think of histamine when thinking of allergies. Indigenous fermented foods and beverages, as potential sources of probiotics, may be very therapeutic for some, while others may have an intolerance to histamine since there is no histamine free diet, and this amine, with many functions in the body, occurs naturally, and is a neurotransmitter in the central nervous system [117].

Fruits and vegetables are easily perishable commodities in Africa due to their high water activity and nutritive values. This phenomenon is more critical in tropical and subtropical countries, whose climates favor the growth of spoilage causing microorganisms. It is in developing settings in Africa, Asia [118], and Latin America [119,120] that perhaps the greatest need for probiotics and fermented foods exist; however, for many reasons, this is not the case [121].

5. One Health Approach and International Organizations

Antimicrobial resistance is the ability of a microorganism (like bacteria, viruses, and some parasites) to stop an antimicrobial (such as antibiotics, antivirals, and antimalarials) from working against it. As a result, standard treatments become ineffective, infections persist, and may spread to others.

Antibiotic resistance is one of the biggest threats to global health, food security, and development today, and can affect anyone, of any age, in any part of the world. Bacteria, not humans or animals, become antibiotic-resistant; the cause for this is mainly the way antibiotics are prescribed and used without sales supervision and medical or veterinarian control. Tackling antibiotic resistance is a high priority for the United Nations' agencies FAO, and OIE, and the WHO, who are leading multiple initiatives and global action plans.

The United Nations (UN) has become the foremost forum in addressing issues that transcend national boundaries and cannot be resolved by any single country acting alone. While conflict resolution and peacekeeping continue to be among its most visible efforts, the UN, along with its specialized agencies, is also engaged in a wide array of activities to improve people's lives around the world—from disaster relief, through education and advancement of women, to peaceful uses of atomic energy.

Despite great successes since 1953, the UN has, in the past and presently, experienced a number of catastrophic failures, such as the war on sustainable development, global energy goals, refugee and climate change policies, famine, poverty, war conflicts, drugs, diseases, security, terrorism, nuclear proliferation, and human rights issues, and it has suffered disappointing setbacks or complete failures in recent decades.

To improve sanitation and drinking water, the UN organizations FAO and OIE, and the WHO have assumed joint responsibility for addressing zoonotic and other diseases of potentially high socio-economic impact. These international UN organizations developed a Tripartite Concept Note (One Health) setting a course of action and proposing a long term framework for global partnerships which is oriented towards the coordination of global activities addressing health hazards and risks at the human-animal-ecosystems crossroads [122].

The One Health European Joint Program (OHEJP) is a European Commission co-funded scientific collaborative research program intended to help prevent and control food-borne and environmental contaminants that affect human health through joint actions on foodborne zoonoses, antimicrobial resistance, and emerging microbiological hazards [123].

Recognizing the health hazards and risks at the human–animal–ecosystems interface is a key element of their assessments, communication, and management. The One Health approach is considered critical for the emergence of antimicrobial drug resistance and on attending prevalent public health concerns, which comprise emerging infectious, parasitic, and zoonotic diseases [124]. Some 60% of human infectious diseases are of animal origin (zoonoses can be caused by bacteria, fungi, mycobacteria, parasites, viruses, and prions); nearly 75% of emerging human infectious diseases in the past three decades originated in animal-borne (even aquatic) diseases/pathogens [125]. Some 80% of such agents can be used for potential bioterrorism and are also pathogens of animal origin [126].

Through strong partnerships with human, animal, environmental health and civil society organizations and professionals, it is considered possible to stimulate advances concerning a safe and secure world with fewer infectious disease threats to human security. But while this UN One Health initiative has proven to be successful from an emerging and infectious disease perspective, its value still needs to be proven in terms of the exchanges and interactions of different microbiomes and elements of microbial communities' transfer among humans, animals, and the environment [127].

6. One Health, Ecosystems and Veterinary Sciences

The development of new technologies to perform DNA sequencing has expanded studies on entire microbial communities in humans, animals, and in the environment. The term "microbiota" encompasses the entire complex ecosystem of gut microorganisms, the bulk of which reside mainly in the colon. The terms "microbiome" or the metagenome of the microbiota comprise all of the genetic material within a microbiota.

A complete understanding of human microbiomes in various body mucosa and surfaces requires an evolutionary perspective. The coevolution of humans and microbiota has generated host-specific microbiome structures and gut homeostasis of physiologic, metabolic, and antigenic diversity [128].

Population growth and economic development are leading to rapid changes in our global ecosystems [129]. Health risks are also a result of broader pressure on ecosystems, from the depletion and degradation of freshwater resources to the impacts of global climate change on natural disasters and agricultural production [130]. There is increased connectivity between humans, domestic pets, wildlife, farm animals, and real-world issues such as sanitation, economics, and food security. Ecosystems, landscapes, and a One Health paradigm, including social-ecological holistic approaches become increasingly important [131,132]. Such interactions require the integration of health science disciplines that span the spectrum from personalized care to public health [133].

At the national and international levels, these domains are organized in different Ministries, and there is an obstruction by professional corporatism which may impair the implementation of the One Health approach by not pursuing to unify health-related research. Furthermore, overcoming long-standing barriers of privacy and distrust among health professionals and political will are necessary to enable the integration of different health systems [134].

Fermentation, as a human ecological process, begins with the symbiotic human relationship with the microbial habitat [135]. Lifestyle, well-being, and even the survival of humans has been connected to single-celled microorganisms, i.e., fungi (yeasts) and bacteria on fermentation ecosystems [136]. The concept of a whole ecosystem is unpopular, and many have abandoned the idea that ecosystems have boundaries [137].

Gut microbes are extensively purged every one to two days and have the ability to double in number within the space of an hour [138]. In future the ecology of human nutrition may be studied on fermentation ecosystems models [139].

The One Health approach has been criticized for an excessive focus on emerging zoonotic diseases, inadequate incorporation of environmental concepts and expertise, and insufficient incorporation of social science and behavioral aspects of health and governance [140]. Barriers to implementing this strategy include competition over budgets, poor communication, and the need for improved technology [141].

At the national level, it is common to observe Veterinary Medicine, Animal and Veterinary Science, Colleges of Veterinary Medicine, and Veterinary Public Health within the Ministry of Agriculture and not in the Ministry of Health; therefore the link between animal health and human health is very precarious. Veterinary medicine is considered an Agrarian profession which does not include the concerns of Health professionals on most criteria, including resources. Environmental health is under the Ministry of Environment, and overall, this partition of responsibilities results in practical difficulties in terms of implementing the collaboration of multiple disciplines and sectors working locally, nationally, and globally to attain optimal health for people, animals, and the environment.

At the international level, agencies such as The European Centre for Disease Prevention and Control (ECDC), The European Food Safety Authority (EFSA), the *forum* One Health European Joint Programme (OHEJP), and others, follow the developments on zoonoses with the mission of identifying, assessing and communicating current and emerging threats to human health posed by these diseases, as well as zoonotic agents, antimicrobial resistance, and food-borne outbreaks. However, the monitoring and surveillance schemes of most zoonotic agents are not harmonized between Member States, adding to existing complexity.

7. Concluding Remarks

Fermented food microbiology is an excellent model that is deeply connected to the dynamics that shape the human microbiota in different body sites. Perceiving microbial community interactions, essential for the threat of global antimicrobial resistance, will help to reveal, via a holistic approach, the unknown secrets of the human microbiome and the interactions which greatly influence multiple forms of human health, nutrition, well-being.

The relevance and potential of fermented foods and beverages, with contrasting and inconclusive results, and advocacy for their inclusion into dietary guidelines, depend on future clinical research. The limitations and inconsistencies in the current body of evidence mean that, presently, no definitive conclusions can be drawn on the potential health benefits of fermented products.

It is not easy to apply trans-inter-multi-disciplinary research required by the One Health approach due to its complexity, but the associated human-animal-environmental microbiota and health threats and risks demand that many challenges and handicaps must be overcome. After two decades, One Health still needs to prove its use and its ability to be applicable in parallel with the present bold reforms which are underway among major United Nations departments in order to more effectively respond to global crises, streamline activities, increase accountability, and ensure effectiveness.

Author Contributions: V.B. & T.F. reviewed the literature and drafted the manuscript. J.F., L.P., M.P. read and revised the manuscript. T.F. responsible for the concept and preparation of final article.

Funding: This research received no external funding.

Conflicts of Interest: The authors declare no conflicts of interest.

References

1. Han, H.; Li, Y.; Fang, J.; Liu, G.; Yin, J.; Li, T.; Yin, Y. Gut Microbiota and Type 1 Diabetes. *Int. J. Mol. Sci.* **2018**, *19*, 995. [CrossRef] [PubMed]
2. Kostic, A.D.; Michael, R.; Howitt, M.R.; Garrett, W.S. Exploring host–microbiota interactions in animal models and humans. *Genes Dev.* **2013**, *27*, 701–718. [CrossRef] [PubMed]
3. Aagaard, K.; Ma, J.; Antony, K.M.; Ganu, R.; Petrosino, J.; Versalovic, J. The Placenta Harbors a Unique Microbiome. *Sci. Transl. Med.* **2014**, *6*, 237ra65. [CrossRef] [PubMed]
4. Walker, R.W.; Clemente, J.C.; Peter, I.; Loos, R.J.F. The prenatal gut microbiome: Are we colonized with bacteria in utero? *Pediatr. Obes.* **2017**, *12* (Suppl. 1), 3–17. [CrossRef] [PubMed]
5. Dzidic, M.; Boix-Amorós, A.; Selma-Royo, M.; Mira, A.; Collado, M.C. Gut Microbiota and Mucosal Immunity in the Neonate. *Med. Sci.* **2018**, *6*, 56. [CrossRef] [PubMed]
6. Perez-Muñoz, M.E.; Arrieta, M.-C.; Ramer-Tait, A.E.; Walter, J. A critical assessment of the "sterile womb" and "in utero colonization" hypotheses: Implications for research on the pioneer infant microbiome. *Microbiome* **2017**, *5*, 48. [CrossRef] [PubMed]
7. O'Hara, A.M.; Shanahan, F. The gut flora as a forgotten organ. *EMBO Rep.* **2006**, *7*, 688–693. [CrossRef]

8. Cenit, M.C.; Sanz, Y.; Codoñer-Franch, P. Influence of gut microbiota on neuropsychiatric disorders. *World J. Gastroenterol.* **2017**, *23*, 5486–5498. [CrossRef]

9. Lira-Junior, R.; Boström, E.A. Oral-gut connection: One step closer to an integrated view of the gastrointestinal tract? *Mucosal Immunol.* **2018**, *11*, 316–318. [CrossRef]

10. Wampach, L.; Heintz-Buschart, A.; Hogan, A.; Muller, E.E.; Narayanasamy, S.; Laczny, C.C.; Hugerth, L.W.; Bindl, L.; Bottu, J.; Andersson, A.F.; et al. Colonization and succession within the human gut microbiome by archaea, bacteria, and microeukaryotes during the first year of life. *Front. Microbiol.* **2017**, *8*, 738. [CrossRef]

11. Sender, R.; Fuchs, S.; Milo, R. Revised Estimates for the Number of Human and Bacteria Cells in the Body. *PLoS Biol.* **2016**, *14*, e1002533. [CrossRef] [PubMed]

12. Belizário, J.E.; Napolitano, M. Human microbiomes and their roles in dysbiosis, common diseases, and novel therapeutic approaches. *Front. Microbiol.* **2015**, *6*, 1050. [CrossRef] [PubMed]

13. Vemuri, R.; Gundamaraju, R.; Shastri, M.D.; Shukla, S.D.; Kalpurath, K.; Ball, M.; Tristram, S.; Shankar, E.M.; Ahuja, K.; Eri, R. Gut Microbial Changes, Interactions, and Their Implications on Human Lifecycle: An Ageing Perspective. *Biomed. Res Int.* **2018**, *2018*, 4178607. [CrossRef] [PubMed]

14. Thursby, E.; Juge, N. Introduction to the human gut microbiota. *Biochem. J.* **2017**, *474*, 1823–1836. [CrossRef] [PubMed]

15. Schloss, P.D.; Handelsman, J. Metagenomics for studying unculturable microorganisms: Cutting the Gordian knot. *Genome Biol.* **2005**, *6*, 229. [CrossRef] [PubMed]

16. Guirro, M.; Costa, A.; Gual-Grau, A.; Mayneris-Perxachs, J.; Torrell, H.; Herrero, P.; Canela, N.; Arola, L. Multi-omics approach to elucidate the gut microbiota activity: Metaproteomics and metagenomics connection. *Electrophoresis* **2018**, *39*, 1692–1701. [CrossRef] [PubMed]

17. Zhang, C.; Zhao, L. Strain-level dissection of the contribution of the gut microbiome to human metabolic disease. *Genome Med.* **2016**, *8*, 41. [CrossRef]

18. Butel, M.J.; Waligora-Dupriet, A.J.; Wydau-Dematteis, S. The developing gut microbiota and its consequences for health. *J. Dev. Orig. Health Dis.* **2018**, 1–8. [CrossRef]

19. Wu, G.; Zhang, C.; Wu, H.; Wang, R.; Shen, J.; Wang, L.; Zhao, Y.; Pang, X.; Zhang, X.; Zhao, L.; et al. Genomic microdiversity of Bifidobacterium pseudocatenulatum underlying differential strain-level responses to dietary carbohydrate intervention. *MBio* **2017**, *8*, e02348-16. [CrossRef]

20. Dahiya, D.K.; Renuka; Puniya, M.; Shandilya, U.K.; Dhewa, T.; Kumar, N.; Kumar, S.; Puniya, A.K.; Shukla, P. Gut Microbiota Modulation and Its Relationship with Obesity Using Prebiotic Fibers and Probiotics: A Review. *Front. Microbiol.* **2017**, *8*, 563. [CrossRef]

21. Gagliardi, A.; Totino, V.; Cacciotti, F.; Iebba, V.; Neroni, B.; Bonfiglio, V.; Trancassini, M.; Passariello, C.; Pantanella, F.; Schippa, S. Rebuilding of the gut ecosystem. *Int. J. Environ. Res. Public Health* **2018**, *15*, 1679. [CrossRef] [PubMed]

22. Cicerone, C.; Nenna, R.; Pontone, S. Th17, intestinal microbiota and the abnormal immune response in the pathogenesis of celiac disease. *Gastroenterol. Hepatol. Bed Bench* **2015**, *8*, 117–122. [PubMed]

23. Hemarajata, P.; Versalovic, J. Effects of probiotics on gut microbiota: Mechanisms of intestinal immunomodulation and neuromodulation. *Therap. Adv. Gastroenterol.* **2013**, *6*, 39–51. [CrossRef] [PubMed]

24. Shimizu, K.; Yamada, T.; Ogura, H.; Mohri, T.; Kiguchi, T.; Fujimi, S.; Asahara, T.; Yamada, T.; Ojima, M.; Ikeda, M.; et al. Synbiotics modulate gut microbiota and reduce enteritis and ventilator-associated pneumonia in patients with sepsis: A randomized controlled trial. *Crit. Care* **2018**, *22*, 239. [CrossRef] [PubMed]

25. Zuo, T.; Ng, S.C. The Gut Microbiota in the Pathogenesis and Therapeutics of Inflammatory Bowel Disease. *Front. Microbiol.* **2018**, *9*, 2247. [CrossRef] [PubMed]

26. Zmora, N.; Zilberman-Schapira, G.; Suez, J.; Mor, U.; Dori-Bachash, M.; Bashiardes, S.; Kotler, E.; Zur, M.; Regev-Lehavi, D.; Brik, R.B.; et al. Personalized Gut Mucosal Colonization Resistance to Empiric Probiotics Is Associated with Unique Host and Microbiome Features. *Cell* **2018**, *174*, 1388–1405. [CrossRef] [PubMed]

27. Gibbs, E.P.J. The evolution of One Health: A decade of progress and challenges for the future. *Vet. Rec.* **2014**, *174*, 85–91. [CrossRef]

28. FAO/OIE/WHO. *WHO Global Principles for the Containment of Antimicrobial Resistance in Animals Intended for Food 2000*; WHO: Geneva, Switzerland, 2000.

29. Stadtländer, C.T.K.-H. One Health: People, animals, and the environment. *Infect. Ecol. Epidemiol.* **2015**, *5*, 30514. [CrossRef]

30. Janakiraman, M.; Krishnamoorthy, G. Emerging Role of Diet and Microbiota Interactions in Neuroinflammation. *Front. Immunol.* **2018**, *9*, 2067. [CrossRef]

31. Hashemi, Z.; Fouhse, J.; Im, H.S.; Chan, C.B.; Willing, B.P. Dietary pea fiber supplementation improves glycemia and induces changes in the composition of gut microbiota, serum short-chain fatty acid profile and expression of mucins in glucose intolerant rats. *Nutrients* **2017**, *9*, 1236. [CrossRef]

32. Rubio, C.; Schmidt, P.T. Severe Defects in the Macrophage Barrier to Gut Microflora in Inflammatory Bowel Disease and Colon Cancer. *Anticancer Res.* **2018**, *38*, 3811–3815. [CrossRef] [PubMed]

33. Gao, J.; Xu, K.; Liu, G.; Liu, G.; Bai, M.; Peng, C.; Li, T.; Yin, Y. Impact of the Gut Microbiota on Intestinal Immunity Mediated by Tryptophan Metabolism. *Front. Cell. Infect. Microbiol.* **2018**, *8*, 13. [CrossRef] [PubMed]

34. Opazo, M.C.; Ortega-Rocha, E.M.; Coronado-Arrázola, I.; Bonifaz, L.C.; Boudin, H.; Neunlist, M.; Bueno, S.M.; Kalergis, A.M.; Riedel, C.A. Intestinal Microbiota Influences Non-intestinal Related Autoimmune Diseases. *Front. Microbiol.* **2018**, *9*, 432. [CrossRef] [PubMed]

35. Jandhyala, S.M.; Talukdar, R.; Subramanyam, C.; Vuyyuru, H.; Sasikala, M.; Reddy, D.N. Role of the normal gut microbiota. *World J. Gastroenterol.* **2015**, *21*, 8787–8803. [CrossRef] [PubMed]

36. Suez, J.; Zmora, N.; Zilberman-Schapira, G.; Mor, U.; Dori-Bachash, M.; Bashiardes, S.; Zur, M.; Regev-Lehavi, D.; Ben-Zeev Brik, R.; Federici, S.; et al. Post-Antibiotic Gut Mucosal Microbiome Reconstitution Is Impaired by Probiotics and Improved by Autologous FMT. *Cell* **2018**, *174*, 1406–1423. [CrossRef]

37. Severyn, C.J.; Bhatt, A.S. With probiotics, resistance is not always futile. *Cell Host Microbe* **2018**, *24*, 334. [CrossRef] [PubMed]

38. Parker, E.A.; Roy, T.; D'Adamo, C.R.; Wieland, L.S. Probiotics and gastrointestinal conditions: An overview of evidence from the Cochrane Collaboration. *Nutrition* **2018**, *45*, 125–134. [CrossRef]

39. Azad, M.A.K.; Sarker, M.; Li, T.; Yin, J. Probiotic Species in the Modulation of Gut Microbiota: An Overview. *Biomed. Res. Int.* **2018**, 9478630. [CrossRef]

40. Fadgyas-Stanculete, M.; Buga, A.-M.; Popa-Wagner, A.; Dumitrascu, D.L. The relationship between irritable bowel syndrome and psychiatric disorders: From molecular changes to clinical manifestations. *J. Mol. Psychiatry* **2014**, *2*, 4. [CrossRef]

41. Lee, P.; Yacyshyn, B.R.; Yacyshyn, M. Gut microbiota and obesity: An opportunity to alter obesity through Fecal Microbiota Transplant (FMT). *Diabetes Obes. Metab.* **2018**. [CrossRef]

42. Karimi, P.; Kamali, E.; Mousavi, S.M.; Karahmadi, M. Environmental factors influencing the risk of autism. *J. Res. Med. Sci.* **2017**, *22*, 27. [CrossRef] [PubMed]

43. Gérard, P. Gut microbiota and obesity. *Cell Mol. Life Sci.* **2016**, *73*, 147–162. [CrossRef] [PubMed]

44. Delhey, L.; Kilinc, E.N.; Yin, L.; Slattery, J.; Tippett, M.; Wynne, R.; Rose, S.; Kahler, S.; Damle, S.; Legido, A.; et al. Bioenergetic variation is related to autism symptomatology. *Metab. Brain Dis.* **2017**, *32*, 2021–2031. [CrossRef] [PubMed]

45. Freedman, S.N.; Shahi, S.K.; Mangalam, A.K. The "Gut Feeling": Breaking Down the Role of Gut Microbiome in Multiple Sclerosis. *Neurotherapeutics* **2018**, *15*, 109–125. [CrossRef] [PubMed]

46. Adams, J.B.; Audhya, T.; Geis, E.; Gehn, E.; Fimbres, V.; Pollard, E.L.; Mitchell, J.; Ingram, J.; Hellmers, R.; Laake, D.; et al. Comprehensive Nutritional and Dietary Intervention for Autism Spectrum Disorder—A Randomized, Controlled 12-Month Trial. *Nutrients* **2018**, *10*, 369. [CrossRef] [PubMed]

47. Hughes, H.K.; Rose, D.; Ashwood, P. The Gut Microbiota and Dysbiosis in Autism Spectrum Disorders. *Curr. Neurol. Neurosci. Rep.* **2018**, *18*, 81. [CrossRef] [PubMed]

48. Mohajeri, M.H.; Brummer, R.J.M.; Rastall, R.A.; Weersma, R.K.; Harmsen, H.J.M.; Faas, M.; Eggersdorfer, M. The role of the microbiome for human health: From basic science to clinical applications. *Eur. J. Nutr.* **2018**, *57* (Suppl. 1), 1–4. [CrossRef]

49. Goldenberg, R.L.; Hauth, J.C.; Andrews, W.W. Intrauterine infection and preterm delivery. *N. Engl. J. Med.* **2000**, *342*, 1500–1507. [CrossRef]

50. Sánchez, B.; Delgado, S.; Blanco-Míguez, A.; Lourenço, A.; Gueimonde, M.; Margolles, A. Probiotics, gut microbiota, and their influence on host health and disease. *Mol. Nutr. Food Res.* **2017**, *61*. [CrossRef]

51. Singh, R.K.; Chang, H.W.; Yan, D.; Lee, K.M.; Ucmak, D.; Wong, K.; Abrouk, M.; Farahnik, B.; Nakamura, M.; Zhu, T.H.; et al. Influence of diet on the gut microbiome and implications for human health. *J. Transl. Med.* **2017**, *15*, 73. [CrossRef]

52. Howarth, G.S.; Wang, H. Role of endogenous microbiota, probiotics and their biological products in human health. *Nutrients* **2013**, *5*, 58–81. [CrossRef] [PubMed]

53. González-Sarrías, A.; Romo-Vaquero, M.; García-Villalba, R.; Cortés-Martín, A.; Selma, M.V.; Espín, J.C. The Endotoxemia Marker Lipopolysaccharide-Binding Protein is Reduced in Overweight-Obese Subjects Consuming Pomegranate Extract by Modulating the Gut Microbiota: A Randomized Clinical Trial. *Mol. Nutr. Food Res.* **2018**, *62*, 11. [CrossRef] [PubMed]

54. Hutkins, R.W. *Microbiology and Technology of Fermented Foods*, 2nd ed.; Hoboken, N.J., Ed.; Wiley-Blackwell: Hoboken, NJ, USA, September 2018; p. 616.

55. Fridman, W.H.; Zitvogel, L.; Sautes-Fridman, C.; Kroemer, G. The immune contexture in cancer prognosis and treatment. *Nat. Rev. Clin. Oncol.* **2017**, *14*, 717–734. [CrossRef] [PubMed]

56. Iwu, M.M. *Food as Medicine: Functional Food Plants of Africa. Series: Functional Foods and Nutraceuticals*, 1st ed.; CRC Press: Boca Raton, FL, USA, 2016.

57. Borresen, E.C.; Henderson, A.J.; Kumar, A.; Weir, T.L.; Ryan, E.P. Fermented foods: Patented approaches and formulations for nutritional supplementation and health promotion. *Recent Pat. Food Nutr. Agric.* **2012**, *4*, 134–140. [CrossRef] [PubMed]

58. Carrera-Quintanar, L.; Roa, R.I.L.; Quintero-Fabián, S.; Sánchez-Sánchez, M.A.; Vizmanos BOrtuño-Sahagún, D. Phytochemicals That Influence Gut Microbiota as Prophylactics and for the Treatment of Obesity and Inflammatory Diseases. *Mediat. Inflamm.* **2018**, *2018*, 9734845. [CrossRef]

59. Gibson, G.; Hutkins, R.; Sanders, M.E.; Prescott, S.L.; Reimer, R.A.; Salminen, S.J.; Scott, K.; Stanton, C.; Swanson, K.S.; Cani, P.D.; et al. Expert consensus document: The International Scientific Association for Probiotics and Prebiotics (ISAPP) consensus statement on the definition and scope of prebiotics. *Nat. Rev. Gastroenterol. Hepatol.* **2017**, *14*, 491–502. [CrossRef] [PubMed]

60. Parker, R.B. Probiotics, the other half of the antibiotic story. *Anim. Nutr. Health.* **1974**, *29*, 4–8.

61. Metchnikoff, I.I. *The Prolongation of Life: Optimistic Studies*; Springer Publishing Company: New York, NY, USA, 2004.

62. Reid, G. Probiotics: Definition, scope and mechanisms of action. *Best Pract. Res. Clin. Gastroenterol.* **2016**, *30*, 17–25. [CrossRef]

63. Gismondo, M.R.; Drago, L.; Lombardi, A. Review of Probiotics available to modify gastrointestinal flora. *Int. J. Antimicrob. Agents* **1999**, *12*, 287–292. [CrossRef]

64. Hill, C.; Guarner, F.; Reid, G.; Gibson, G.R.; Merenstein, D.J.; Pot, B.; Morelli, L.; Canani, R.B.; Flint, H.J.; Salminen, S.; et al. The International Scientific Association for Probiotics and Prebiotics consensus statement on the scope and appropriate use of the term probiotic. *Nat. Rev. Gastroenterol. Hepatol.* **2014**, *11*, 506–514. [CrossRef]

65. Reid, G.; Sanders, M.E.; Gaskins, H.R.; Gibson, G.R.; Mercenier, A.; Rastall, R.; Roberfroid, M.; Rowland, I.; Cherbut, C.; Klaenhammer, T.R. New scientific paradigms for Probiotics and prebiotics. *J. Clin. Gastroenterol.* **2003**, *37*, 105–118. [CrossRef] [PubMed]

66. Brusaferro, A.; Cavalli, E.; Farinelli, E.; Cozzali, R.; Principi, N.; Esposito, S. Gut dysbiosis and paediatric crohn's disease. *J. Infect.* **2018**. [CrossRef] [PubMed]

67. Bell, V.; Ferrão, J.; Fernandes, T. Nutritional Guidelines and Fermented Food Frameworks. *Foods* **2017**, *6*, 65. [CrossRef]

68. Gille, D.; Schmid, A.; Walther, B.; Vergères, G. Fermented Food and Non-Communicable Chronic Diseases: A Review. *Nutrients* **2018**, *10*, 448. [CrossRef] [PubMed]

69. Marco, M.L.; Heeney, D.; Binda, S.; Cifelli, C.J.; Cotter, P.D.; Foligné, B.; Gänzle, M.; Kort, R.; Pasin, G.; Pihlanto, A.; et al. Health benefits of fermented foods: Microbiota and beyond. *Curr. Opin. Biotechnol.* **2017**, *44*, 94–102. [CrossRef] [PubMed]

70. Hoffmann, D.E.; Fraser, C.M.; Palumbo, F.B.; Ravel, J.; Rothenberg, K.; Rowthorn, V.; Schwartz, J. Science and regulation. Probiotics: Finding the right regulatory balance. *Science* **2013**, *342*, 314–315. [CrossRef] [PubMed]

71. Amit, S.K.; Uddin, M.; Rahman, R.; Islam, R.; Khan, M.S. A review on mechanisms and commercial aspects of food preservation and processing. *Agric. Food Secur.* **2017**, *6*, 51. [CrossRef]

72. Sivamaruthi, B.S.; Kesika, P.; Chaiyasut, C. Impact of Fermented Foods on Human Cognitive Function—A Review of Outcome of Clinical Trials. *Sci. Pharm.* **2018**, *86*, 22. [CrossRef]

73. European Food Safety Authority (EFSA) Panel on Dietetic Products, Nutrition and Allergies. Scientific Opinion on the substantiation of health claims related to live yoghurt cultures and improved lactose digestion (ID 1143, 2976) pursuant to Article 13(1) of Regulation (EC) No 1924/2006. *EFSA J.* **2010**, *8*, 1763.

74. British Nutrition Foundation. Available online: https://www.nutrition.org.uk/nutritionscience/foodfacts/functional-foods.html (accessed on 11 October 2018).

75. Hu, S.; Cao, X.; Wu, Y.; Mei, X.; Xu, H.; Wang, Y.; Zhang, X.; Gong, L.; Li, W. Effects of Probiotic Bacillus as an Alternative of Antibiotics on Digestive Enzymes Activity and Intestinal Integrity of Piglets. *Front. Microbiol.* **2018**, *9*, 2427. [CrossRef]

76. Laurent-Babot, C.; Guyot, J.P. Should Research on the Nutritional Potential and Health Benefits of Fermented Cereals Focus More on the General Health Status of Populations in Developing Countries? *Microorganisms* **2017**, *5*, 40. [CrossRef] [PubMed]

77. Rezac, S.; Kok, C.R.; Heermann, M.; Hutkins, R. Fermented Foods as a Dietary Source of Live Organisms. *Front. Microbiol.* **2018**, *9*, 1785. [CrossRef] [PubMed]

78. Behravesh, C.B. One Health: People, Animals, and the Environment. *Emerg. Infect. Dis.* **2016**, *22*, 766–767. [CrossRef]

79. Kim, N.; Yun, M.; Oh, Y.J.; Choi, H.J. Mind-altering with the gut: Modulation of the gut-brain axis with probiotics. *J. Microbiol.* **2018**, *56*, 172–182. [CrossRef] [PubMed]

80. Dicks, L.M.; Dreyer, L.; Smith, C.; Van Staden, A.D. A Review: The Fate of Bacteriocins in the Human Gastro-Intestinal Tract: Do They Cross the Gut-Blood Barrier? *Front. Microbiol.* **2018**, *9*, 2297. [CrossRef] [PubMed]

81. Ferrão, J.; Bell, V.; Calabrese, V.; Pimentel, L.; Pintado, M.; Fernandes, T.H. Impact of Mushroom Nutrition on Microbiota and Potential for Preventative Health. *J. Food Nutr. Res.* **2017**, *5*, 226–233. [CrossRef]

82. Vamanu, E.; Pelinescu, D.; Sarbu, I. Comparative Fingerprinting of the Human Microbiota in Diabetes and Cardiovascular Disease. *J. Med. Food* **2016**, *19*, 1188–1195. [CrossRef]

83. Gianfranceschi, G.L.; Gianfranceschi, G.; Quassinti, L.; Bramucci, M. Biochemical requirements of bioactive peptides for nutraceutical efficacy. *J. Funct. Foods* **2018**, *47*, 252–263. [CrossRef]

84. Park, Y.W.; Nam, M.S. Bioactive Peptides in Milk and Dairy Products: A Review. *Korean J. Food Sci. Anim. Resour.* **2015**, *35*, 831–840. [CrossRef]

85. Hayes, M.; García-Vaquero, M. Bioactive Compounds from Fermented Food Products. In *Novel Food Fermentation Technologies*; Ojha, K., Tiwari, B., Eds.; Food Engineering Series; Springer: Cham, Switzerland, 2016.

86. Penna, M.; Gębski, J.; Gutkowska, K.; Żakowska-Biemans, S. Importance of Health Aspects in Polish Consumer Choices of Dairy Products. *Nutrients* **2018**, *10*, 1007. [CrossRef]

87. Annunziata, A.; Mariani, A. Consumer Perception of Sustainability Attributes in Organic and Local Food. *Recent Pat. Food Nutr. Agric.* **2018**, *9*, 87–96. [CrossRef]

88. Cryan, J.F.; O'Mahony, S.M. The microbiome-gut-brain axis: From bowel to behavior. *Neurogastroenterol. Motil.* **2011**, *23*, 187–192. [CrossRef] [PubMed]

89. Browne, H.P.; Neville, B.A.; Forster, S.C.; Lawley, T.D. Transmission of the gut microbiota: Spreading of health. *Nat. Rev. Microbiol.* **2017**, *15*, 531–543. [CrossRef] [PubMed]

90. Navarro, F.; Liu, Y.; Rhoads, J.M. Can probiotics benefit children with autism spectrum disorders? *World J. Gastroenterol.* **2016**, *22*, 10093–10102. [CrossRef] [PubMed]

91. Gilbert, J.A.; Krajmalnik-Brown, R.; Porazinska, D.L.; Weiss, S.J.; Knight, R. Toward Effective Probiotics for Autism and Other Neurodevelopmental Disorders. *Cell* **2013**, *155*, 1446–1448. [CrossRef] [PubMed]

92. Liang, S.; Wu, X.; Jin, F. Gut-Brain Psychology: Rethinking Psychology from the Microbiota-Gut-Brain Axis. *Front. Integr. Neurosci.* **2018**, *12*, 33. [CrossRef] [PubMed]

93. Yang, Y.; Tian, J.; Yang, B. Targeting gut microbiome: A novel and potential therapy for autism. *Life Sci.* **2018**, *194*, 111–119. [CrossRef]

94. Genser, L.; Aguanno, D.; Soula, H.A.; Dong, L.; Trystram, L.; Assmann, K.; Salem, J.-E.; Vaillant, J.-C.; Oppert, J.M.; Laugerette, F.; et al. Increased jejunal permeability in human obesity is revealed by a lipid challenge and is linked to inflammation and type 2 diabetes: Jejunal permeability in human obesity. *J. Pathol.* **2018**, *246*. [CrossRef]

95. Hiippala, K.; Jouhten, H.; Ronk ainen, A.; Hartikainen, A.; Kailunanen, V.; Jalanka, J.; Satokari, R. The potential of gut commensals in reinforcing intestinal barrier function and alleviating inflammation. *Nutrients* **2018**, *10*, 988. [CrossRef]

96. Wang, L.W.; Tancredi, D.J.; Thomas, D.W. The prevalence of gastrointestinal problems in children across the United States with autism spectrum disorders from families with multiple affected members. *J. Dev. Behav. Pediatr.* **2011**, *32*, 351–360. [CrossRef]

97. Heidari, F.; Abbaszadeh, S.; Mirak, S.E.M. Evaluation Effect of Combination Probiotics and Antibiotics in the Prevention of Recurrent Urinary Tract Infection (UTI) in Women. *Biomed. Pharmacol. J.* **2017**, *10*, 691–698. [CrossRef]

98. Duca, F.; Gérard, P.; Covasa, M.; Lepage, P. Metabolic interplay between gut bacteria and their host. *Front. Horm. Res.* **2014**, *42*, 73–82. [CrossRef]

99. Strati, F.; Cavalieri, D.; Albanese, D.; De Felice, C.; Donati, C.; Hayek, J.; Jousson, O.; Leoncini, S.; Renzi, D.; Calabrò, A.; et al. New evidences on the altered gut microbiota in autism spectrum disorders. *Microbiome* **2017**, *5*, 24. [CrossRef] [PubMed]

100. Moschen, A.R.; Wieser, V.; Tilg, H. Dietary factors: Major regulators of the gut's microbiota. *Gut Liver* **2012**, *6*, 411–416. [CrossRef] [PubMed]

101. Piwowarczyk, A.; Horvath, A.; Łukasik, J.; Pisula, E.; Szajewska, H. Gluten- and casein-free diet and autism spectrum disorders in children: A systematic review. *Eur. J. Nutr.* **2018**, *57*, 433–440. [CrossRef] [PubMed]

102. Kho, Z.Y.; Lal, S.K. The Human Gut Microbiome—A Potential Controller of Wellness and Disease. *Front. Microbiol.* **2018**, *9*, 1835. [CrossRef] [PubMed]

103. Miller, A.H.; Raison, C.L. The role of inflammation in depression: From evolutionary imperative to modern treatment target. *Nat. Rev. Immunol.* **2015**, *16*, 22–34. [CrossRef]

104. Vuong, H.E.; Hsiao, E.Y. Emerging Roles for the Gut Microbiome in Autism Spectrum Disorder. *Rev. Art. Biol. Psychiatry* **2017**, *81*, 411–423. [CrossRef]

105. Berding, K.; Donovan, S.M. Microbiome and nutrition in autism spectrum disorder: Current knowledge and research needs. *Nutr. Rev.* **2016**, *74*, 723–736. [CrossRef]

106. Den Besten, H.M.W.; Amézquita, A.; Bover-Cid, S.; Dagnas, S.; Ellouze, M.; Guillou, S.; Nychas, G.; O'Mahony, C.; Pérez-Rodriguez, F.; Membré, J.-M. Next generation of microbiological risk assessment: Potential of omics data for exposure assessment. *Int. J. Food Microbiol.* **2018**, *287*, 18–27. [CrossRef]

107. Sirangelo, T.M. Human Gut Microbiome Analysis and Multi-omics Approach. *Int. J. Pharma Med. Biol. Sci.* **2018**, *7*, 52–57.

108. He, M.; Shi, B. Gut microbiota as a potential target of metabolic syndrome: The role of probiotics and prebiotics. *Cell Biosci.* **2017**, *7*, 54. [CrossRef] [PubMed]

109. Willson, K.; Situ, C. Systematic Review on Effects of Diet on Gut Microbiota in Relation to Metabolic Syndromes. *J. Clin. Nutr. Metab.* **2017**, *1*, 2.

110. Routy, B.; Gopalakrishnan, V.; Daillère, R.; Zitvogel, L.; Wargo, J.A.; Kroemer, G. The gut microbiota influences anticancer immunosurveillance and general health. *Nat. Rev. Clin. Oncol.* **2018**, *15*, 382–396. [CrossRef] [PubMed]

111. Asare, M.; Danquah, S.A. The relationship between physical activity, sedentary behaviour and mental health in Ghanaian adolescents. *Child Adol. Psychiatry Ment. Health* **2015**, *9*, 11. [CrossRef] [PubMed]

112. Mahomoodally, M.F. Traditional Medicines in Africa: An Appraisal of Ten Potent African Medicinal Plants. *Evid. Based Complement. Altern. Med.* **2013**, *14*, 617459. [CrossRef] [PubMed]

113. Anukam, K.C.; Reid, G. African traditional fermented foods and probiotics. *J. Med. Food* **2009**, *12*, 1177–1184. [CrossRef]

114. Duthé, G.; Rossier, C.; Bonnet, D.; Soura, A.B.; Corker, J. Mental health and urban living in sub-Saharan Africa: Major depressive episodes among the urban poor in Ouagadougou, Burkina Faso. *Popul. Health Metr.* **2016**, *14*, 18. [CrossRef]

115. Simango, C. Potential use of traditional fermented foods for weaning in Zimbabwe. *Soc. Sci. Med.* **1997**, *44*, 1065–1068. [CrossRef]

116. Mugula, J.K.; Sørhaug, T.; Stepaniak, L. Proteolytic activities in togwa, a Tanzanian fermented food. *Int. J. Food Microbiol.* **2003**, *84*, 1–12. [CrossRef]

117. Kohn, J.B. Is there a diet for histamine intolerance? *J. Acad. Nutr. Diet.* **2014**, *114*, 1860. [CrossRef] [PubMed]

118. Swain, M.R.; Anandharaj, M.; Ray, R.C.; Parveen Rani, R. Fermented fruits and vegetables of Asia: A potential source of probiotics. *Biotechnol. Res. Int.* **2014**, *2014*, 250424. [CrossRef] [PubMed]

119. Penna, A.L.B.; Nero, L.A.; Todorov, S.D. Fermented Foods of Latin America: From Traditional Knowledge to Innovative Applications. In *Food Biology Series*; CRC Press: Boca Raton, FL, USA, 2017; p. 338, ISBN 9781498738118.

120. Ramos, C.L.; Schwan, R.F. Technological and nutritional aspects of indigenous Latin America fermented foods. *Curr. Opin. Food Sci.* **2017**, *13*, 97–102. [CrossRef]

121. Franz, C.M.A.P.; Huch, M.; Mathara, J.M.; Abriouel, H.; Benomar, N.; Reid, G.; Galvez, A.; Holzapfel, W.H. African fermented foods and probiotics. *Int. J. Food Microbiol.* **2014**, *190*, 84–96. [CrossRef] [PubMed]

122. Whiteford, H.A.; Degenhardt, L.; Rehm, J.; Baxter, A.J.; Ferrari, A.J.; Erskine, H.E.; Charlson, F.J.; Norman, R.E.; Flaxman, A.D.; Johns, N.; et al. VosT Global burden of disease attributable to mental and substance use disorders: Findings from the Global Burden of Disease Study 2010. *Lancet* **2013**, *382*, 1575–1586. [CrossRef]

123. Trinh, P.; Zaneveld, J.R.; Safranek, S.; Rabinowitz, P.M. One Health Relationships between Human, Animal, and Environmental Microbiomes: A Mini-Review. *Front. Public Health* **2018**, *6*, 235. [CrossRef] [PubMed]

124. Rostal, M.K.; Ross, N.; Machalaba, C.; Cordel, C.; Paweska, J.T.; Karesha, W.B. Benefits of a one health approach: An example using Rift Valley fever. *One Health* **2018**, *5*, 34–36. [CrossRef]

125. Center for Disease Control (CDC). *Emerging Infectious Diseases*; National Center for Infectious Diseases, Centers for Disease Control and Prevention (CDC): Atlanta, GA, USA, 1995.

126. Malan-Muller, S.; Valles-Colomer, M.; Raes, J.; Lowry, C.A.; Seedat, S.; Hemmings, S.M.J. The Gut Microbiome and Mental Health: Implications for Anxiety- and Trauma-Related Disorders. *OMICS* **2018**, *22*, 90–107. [CrossRef]

127. El Samra, G.H. Climate change, food security, food safety and nutrition. *Egypt. J. Occup. Med.* **2017**, *41*, 217–236.

128. Wells, J.M.; Brummer, R.J.; Derrien, M.; MacDonald, T.T.; Troost, F.; Cani, P.D.; Theodorou, V.; Dekker, J.; Méheust, A.; de Vos, W.M.; et al. Homeostasis of the gut barrier and potential biomarkers. *Am. J. Physiol. Gastrointest. Liver Physiol.* **2016**, *312*, G171–G193. [CrossRef]

129. Galvani, A.P.; Bauch, C.T.; Anand, M.; Singer, B.H.; Levin, S.A. Interactions in population and ecosystem health. *Proc. Natl. Acad. Sci. USA* **2016**, *113*, 14502–14506. [CrossRef] [PubMed]

130. Ecosystems and Human Well-Being. Available online: http://apps.who.int/iris/bitstream/handle/10665/43354/9241563095.pdf?sequence=1 (accessed on 30 November 2018).

131. Davenport, E.R.; Sanders, J.G.; Song, S.J.; Amato, K.R.; Clark, A.G.; Knight, R. The human microbiome in evolution. *BMC Biol.* **2017**, *15*, 127. [CrossRef] [PubMed]

132. Fischer, A.P. Forest landscapes as social-ecological systems and implications for management. *Landsc. Urban Plan.* **2018**, *177*, 138–147. [CrossRef]

133. Murtaugh, M.P.; Steer, C.J.; Sreevatsan, S.; Patterson, N.; Kennedy, S.; Sriramarao, P. The science behind One Health: At the interface of humans, animals, and the environment. *Ann. N. Y. Acad. Sci.* **2017**, *1395*, 12–32. [CrossRef] [PubMed]

134. Johnson, I.; Hansen, A.; Bi, P. The challenges of implementing an integrated One Health surveillance system in Australia. *Zoonoses Public Health* **2018**, *65*, e229–e236. [CrossRef] [PubMed]

135. Wu, H.-J.; Wu, E. The role of gut microbiota in immune homeostasis and autoimmunity. *Gut Microb.* **2012**, *3*, 4–14. [CrossRef] [PubMed]

136. Zhao, Y.; Yu, Y.-B. Intestinal microbiota and chronic constipation. *SpringerPlus* **2016**, *5*, 1130. [CrossRef] [PubMed]

137. Atlas, R.; Maloy, S. *One Health: People, Animals, and the Environment*; ASM Press: Washington, DC, USA, 2014; ISBN 978-1555818425. [CrossRef]

138. David, L.A.; Maurice, C.F.; Carmody, R.N.; Gootenberg, D.B.; Button, J.E.; Wolfe, B.E.; Ling, A.V.; Devlin, A.S.; Varma, Y.; Fischbach, M.A.; et al. Diet rapidly and reproducibly alters the human gut microbiome. *Nature* **2014**, *505*, 559–563. [CrossRef]

139. Katsnelson, A. Fermented foods offer up microbial model system. *Proc. Natl. Acad. Sci. USA* **2017**, *114*, 2434–2436. [CrossRef]

140. Rabinowitz, P.M.; Pappaioanou, M.; Bardosh, K.L.; Conti, L. A planetary vision for one health. *BMJ Glob. Health* **2018**, *3*, e001137. [CrossRef]
141. Nyatanyi, T.; Wilkes, M.; McDermott, H.; Nzietchueng, S.; Gafarasi, I.; Mudakikwa, A.; Kinani, J.F.; Rukelibuga, J.; Omolo, J.; Mupfasoni, D.; et al. Implementing One Health as an integrated approach to health in Rwanda. *BMJ Glob. Health* **2017**, *2*, e000121. [CrossRef] [PubMed]

 *foods*

Article

Microbial Community Analysis of Sauerkraut Fermentation Reveals a Stable and Rapidly Established Community

Michelle A. Zabat [†], William H. Sano [†], Jenna I. Wurster, Damien J. Cabral and Peter Belenky *

Department of Molecular Microbiology and Immunology, Division of Biology and Medicine, Brown University, Providence, RI 02912, USA; michelle_zabat@brown.edu (M.A.Z.); william_sano@brown.edu (W.H.S.); jenna_wurster@brown.edu (J.I.W.); damien_cabral@brown.edu (D.J.C.)
* Correspondence: peter_belenky@brown.edu; Tel.: +1-401-863-5953
† The two authors contributed equally.

Received: 10 April 2018; Accepted: 10 May 2018; Published: 12 May 2018

Abstract: Despite recent interest in microbial communities of fermented foods, there has been little inquiry into the bacterial community dynamics of sauerkraut, one of the world's oldest and most prevalent fermented foods. In this study, we utilize 16S rRNA amplicon sequencing to profile the microbial community of naturally fermented sauerkraut throughout the fermentation process while also analyzing the bacterial communities of the starting ingredients and the production environment. Our results indicate that the sauerkraut microbiome is rapidly established after fermentation begins and that the community is stable through fermentation and packaging for commercial sale. Our high-throughput analysis is in agreement with previous studies that utilized traditional microbiological assessments but expands the identified taxonomy. Additionally, we find that the microbial communities of the starting ingredients and the production facility environment exhibit low relative abundance of the lactic acid bacteria that dominate fermented sauerkraut.

Keywords: sauerkraut; microbiome; fermentation; probiotics; high-throughput sequencing

---

1. Introduction

Sauerkraut, a fermented food made primarily from cabbage, is one of the most well-known varieties of fermented food, dating back to the Roman Empire. Historically, it served as a source of nutrients during the winter months when fresh food was scarce, as proper fermentation preserves the nutritive value of cabbage while creating desirable sensory properties [1,2]. It is most commonly associated with Central and Eastern European cultures, though it can be found in Western European cuisine as well. Sauerkraut is thought to have been part of the American diet since the country's founding, usually as a cooking ingredient, side dish, or condiment. Its popularity declined beginning in the 1930s as a result of shifting consumer preferences and a lack of product uniformity [1,3]; however, advances in food fermentation science and modern consumer interests have brought sauerkraut renewed popularity in recent years. Today, both mass-produced and artisanal preparations of sauerkraut are widely sold in the United States.

Sauerkraut production and characteristics are largely dependent on the resident microbial community and the fermentation conditions [4]. Though the microbial composition of sauerkraut can vary during the initial stages of fermentation, appropriate fermentation conditions such as temperature and relative ingredient concentration ensure that lactic acid bacteria (LAB) are the dominant microorganisms in the final fermented product. These LAB are of critical importance for successful fermentation; they produce the organic acids, bacteriocins, vitamins, and flavor compounds responsible for many of the characteristic sensory qualities of fermented foods, including extended shelf life, flavor, and nutritional content [5–8]. Additionally, certain LAB have been purported to act as probiotics that contribute to human health and microbiome stability [9,10]. Though these claims have not yet been fully substantiated by scientists, this perspective has contributed to recent increased consumer popularity and consumption in the United States [11].

Canonical sauerkraut fermentation begins with the initial proliferation of *Leuconostoc mesenteroides*, which rapidly produces carbon dioxide and acid. This quickly lowers the environmental pH, inhibiting the growth of undesirable microorganisms that might cause food spoilage while preserving the color of the cabbage [12]. The action of *L. mesenteroides* changes the fermentation environment so that it favors the succession of other LAB, such as *Lactobacillus brevis* and *Lactobacillus plantarum* [12]. In traditional sauerkraut production, this process proceeds at 18 °C for roughly one month [12]. The combination of metabolites that these organisms produce leads to favorable sensory qualities—the unique flavors, aromas, and textures associated with fermented foods—in the final product [12,13]. The temperature of fermentation also plays an important role in terms of color, flavor, and preservability [12].

Historically, the important species in sauerkraut fermentation were considered to be *L. mesenteroides*, *L. plantarum, and L. brevis*, which is supported by recent studies [12,14]. In the event of abnormally high heat or salinity, *Enterococcus faecalis* and *Pediococcus cerevisiae* are thought to play a role in the fermentation process [12]. However, these observations were drawn from studies that used culture-based techniques to isolate bacteria, which are inherently biased due to their inability to capture the range of non-culturable bacteria. Recent studies have also identified the genus *Weissella* as important to early fermentative processes [14].

Recent advances in high-throughput sequencing technology have created the potential for highly accurate, culture-independent characterization of the sauerkraut microbiome. The advent of 16S rRNA amplicon sequencing technology has made it possible to systematically analyze the sauerkraut microbiome before, during, and after fermentation. Sauerkraut fermented at warmer temperatures has historically been considered to be of lower quality than sauerkraut fermented at low temperatures; however, current methods of industrial production are turning towards warm-temperature fermentation because it dramatically shortens production time.

Here, we analyze the taxonomic composition of sauerkraut fermented at room temperature over a 14-day fermentation period. Overall, the taxonomic composition of this sauerkraut is in line with the taxonomic composition observed in sauerkraut fermented in the traditional cold temperature range, suggesting that warm-temperature fermentation may be a viable option for producing a sauerkraut with a bacterial community structure that is in line with sauerkraut produced by a more traditional cold-temperature fermentation. This may be of particular interest to industrial and commercial producers, who would be able to speed their production process without sacrificing the taxonomic composition that is at the heart of consumer interest in probiotics and fermented foods.

2. Materials and Methods

2.1. Sauerkraut Preparation and Sampling Methods

Sauerkraut for this study was sampled from a single 50 lb batch prepared for commercial sale during June 2017 in a facility located near Providence, Rhode Island. Cabbage was salted to a concentration of 2.25% before the addition of caraway seeds (<1% by weight). Ingredient samples were collected in triplicate during a normal production run; 0.5 g of each ingredient were placed into 1.5 mL Eppendorf tubes containing 500 µL of nuclease-free water. The batch of sauerkraut was sealed into airtight plastic drums for the fermentation period. Fermentation was conducted at approximately 21 °C. Successful fermentation was determined by a final pH below 3.6. Fermentation samples were collected in triplicate using Pasteur pipettes from the fermenting sauerkraut at Days 0, 2, 7, 10, and 14. Samples are not true biological replicates, since all triplicates came from the same batch of fermenting sauerkraut; this is a limitation of our study, and future studies should examine the consistency of microbiome dynamics between batches. Packaged, jarred sauerkraut from this producer was purchased from a commercial distributor and processed alongside fermentation samples for microbiome analysis of the finished product.

To sample the production environment, the production table, the industrial sink, and the floor of the production facility were swabbed in triplicate with flocked sterile swabs; these were then stored individually in Zymo Research DNA/RNA Shield Lysis Tubes (Zymo Research, Irvine, CA, USA; Cat: R1103). To sample the air in the facility, empty Petri dishes were left uncovered around the facility throughout the duration of the fermentation period. On Day 14, the Petri dishes were swabbed in the manner described above. After collection, all samples were immediately transported to the laboratory on ice and stored at $-80\,^{\circ}$C until processing.

2.2. DNA Extraction, 16S Library Preparation, and Sequencing

The sauerkraut, environmental, and ingredient samples were processed using the ZymoBIOMICS DNA Microprep Kit (Zymo Research, Irvine, CA, USA; Cat: D4305) according to the manufacturer's instructions in order to extract DNA. Using the Earth Microbiome Project 16S Illumina Amplicon Protocol, we targeted the V4 hypervariable region of the bacterial 16S rRNA gene using an 806Rb reverse primer (GGACTACCAGGGTATCTAATCC) and a barcoded 515F forward primer (CAGCAGCCGCGGTAAT) [15–19]. PCR amplicons were generated using Phusion High-Fidelity polymerase (New England BioLabs, Ipswich, MA, USA) under the following conditions: $98\,^{\circ}$C for 3 minutes, followed by 35 cycles of $98\,^{\circ}$C for 45 s, $50\,^{\circ}$C for 60 s, and $72\,^{\circ}$C for 90 s, and ending with a final elongation at $72\,^{\circ}$C for 10 minutes.

PCR amplicon concentrations were analyzed using the Qubit 3.0 Fluorometer and the dsDNA-HS kit (Thermo Fisher Scientific, Waltham, MA, USA) according to the manufacturer's instructions. Equal amounts of amplicons from each sample were pooled, concentrated, and gel purified using the Machery-Nagel NucleoSpin Gel and PCR Clean-Up kit (Machery-Nagel, Düren, Germany, Cat: 740609) according to the manufacturer's instructions. The pooled samples were submitted to the Rhode Island Genomics and Sequencing Center at the University of Rhode Island (Kingston, RI, USA) for quality control and sequencing. Amplicons were paired-end sequenced (2×250 bp) on an Illumina MiSeq platform using a 500-cycle kit with standard protocols.

2.3. Rarefaction and Sequencing Analysis

The raw paired-end FASTQ reads were demultiplexed using idemp (https://github.com/yhwu/idemp/blob/master/idemp.cp) and imported into the Quantitative Insights Into Microbial Ecology 2 program (QIIME2, ver. 2017.9.0, https://qiime2.org/). Raw reads were subsequently deposited into the National Center for Biotechnology Information (NCBI) Sequence Read Archive (SRA) database under the SRA accession SRP145097. The Divisive Amplicon Denoising Algorithm 2 (DADA2) was used to quality filter, trim, de-noise, and merge the data. Chimeric sequences were removed using the consensus method. A feature classifier in QIIME2 trained with the SILVA 99% operational taxonomic unit (OTU) database and trimmed to the V4 region of the 16S was used to assign taxonomy to all ribosomal sequence variants. Contaminating mitochondrial and chloroplast sequences were filtered out of the resulting feature table. The remaining representative sequences were aligned with MAFFT and used for phylogenetic reconstruction in FastTree. Finally, diversity metrics were calculated using the QIIME2 diversity plugin and visualized with Prism (ver. 7.0a, GraphPad, La Jolla, CA, USA).

After quality filtering and preprocessing, we determined that 8 of our 37 sequenced samples had fewer than 650 reads, which we deemed insufficient for statistically powerful diversity analysis, and thus a potential source of bias. We therefore removed these read-poor samples from downstream alpha and beta diversity analysis. Five of the discarded samples were distributed, one each, across different ingredient and environmental sample types. Given low variation between the remaining two replicates in these sample types, we feel the two replicates are sufficient for publication. The other three read-poor samples were the triplicate fermenting samples from Day 0. These samples were dominated by contaminating chloroplast reads, which were computationally removed. The remaining bacterial reads were sufficiently low that they presented a problem for alpha and beta diversity measurements. The low abundance of bacterial reads in Day 0 samples and our other samples is likely a reflection of the intrinsically low bacterial abundance of those communities. While this does limit the potential scope of our conclusions, it is an inevitable result of working with low abundance communities. This is reflected by the absence of Day 0 in Figures 1 and 2. To visualize the bacterial community at the Day 0 time point, we used a less restrictive cutoff for sample inclusion in our taxa bar plots—250 reads (Figure 3). This allowed us to recapture all three replicates from Day 0 and gain insight into the structure of these communities in the absence of diversity analysis.

3. Results

Alpha diversity values of the environment, ingredients, and fermenting sauerkraut were measured by both the Shannon diversity index and Faith's phylogenetic diversity. These values reveal a reduction in bacterial diversity for fermenting sauerkraut as compared to the starting ingredients and environment (Figure 1). However, we observe contrasting alpha diversity patterns between the Shannon diversity and the Faith's phylogenetic diversity metrics during the fermentation process. The Shannon diversity index indicates a successive increase in alpha diversity of sauerkraut over time while Faith's phylogenetic diversity suggests a constant low alpha diversity. We attribute this to the fact that the Shannon diversity index segments closely related and possibly overlapping LAB into separate taxa. This generates a false appearance of diversity. By contrast, Faith's phylogenetic diversity index uses branch lengths as the basis for assigning diversity metrics, and does not separate LAB with the same level of granularity. Nevertheless, the low level of diversity in the fermenting product shown by both plots is likely the result of selective pressures in the fermentation environment, including low pH, anaerobic conditions, and high salinity. This indicates successful fermentation.

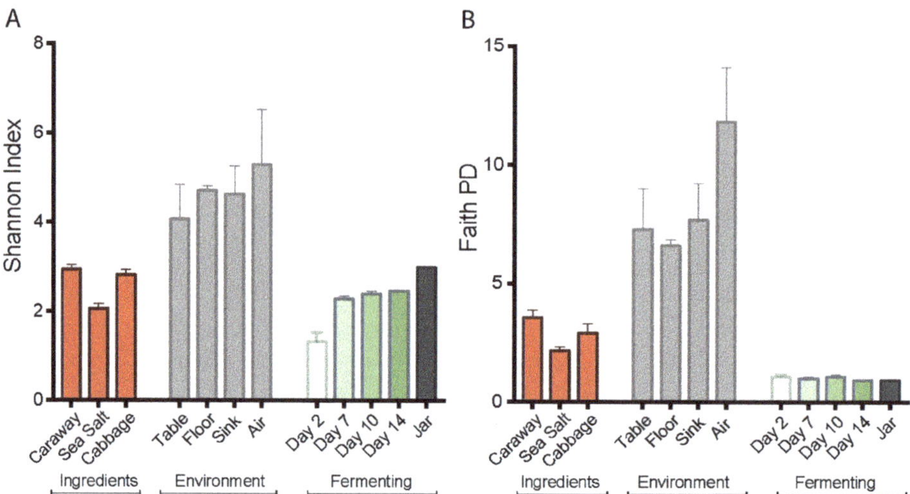

Figure 1. Alpha diversity measures of the sauerkraut, ingredient, and environment samples. (**A**) Shannon index; and (**B**) Faith's phylogenetic diversity (PD). Error bars represent standard error of the mean.

Next we employed principal coordinates analysis (PCoA), with the unweighted UniFrac distance, to visualize the differences in community structures between our samples (Figure 2). The samples exhibited clear clustering by sample type—raw ingredient, environment, or fermentation time point—indicating that the bacterial communities present have significant variation from one another. This is expected for the environmental samples versus raw ingredients or fermentation time points, but was surprising for the raw ingredients versus the fermenting sauerkraut. The Day 0 sample derived from the initial ingredient mixture is much more similar to the Day 14 sauerkraut community than it is to the raw ingredients (Figure 2). This suggests that the selective pressures intrinsic to fermentation have strong and immediate impacts on the bacteria found on and in the raw ingredients.

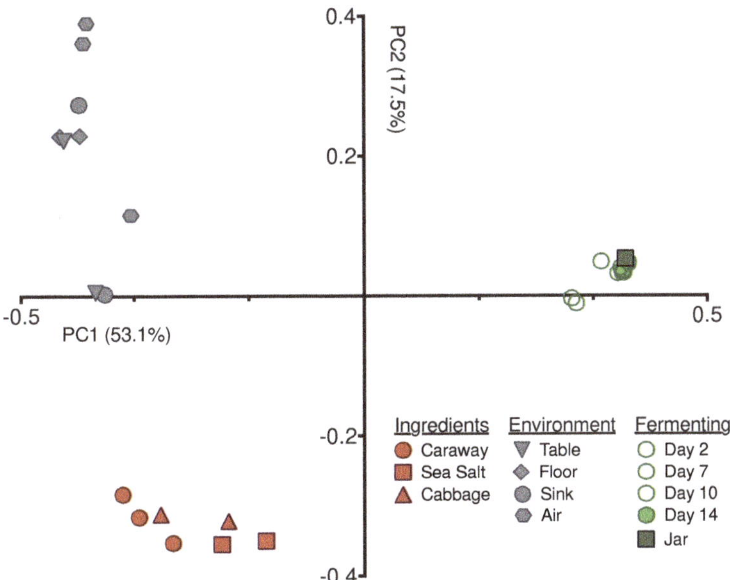

Figure 2. PCoA depicting the unweighted UniFrac distance between fermenting sauerkraut, environment, and ingredient microbiome samples.

To further characterize the bacterial community found in each of the collected samples, we examined the taxonomic structures of our bacterial communities at the order and genus levels (Figure 3). At the order level, we see differences in bacterial composition between the raw ingredient samples, the environmental samples, and the fermentation samples (Figure 3A). The raw ingredient and environmental site categorizations are fairly similar to each other. Sea salt differs slightly from the cabbage and caraway seed, which may be because salt is not a plant product and it likely presents a strong halophilic selection pressure. Overall, these two categories of samples are markedly different from the bacterial communities found during fermentation. The Day 0 sample contains significantly more bacterial taxa than the subsequent time point samples, and illustrates a precipitous drop in the number of bacterial species present over the first 48 hours of fermentation. The most abundant bacterial order in the Day 0 fermentation sample is Pseudomonadales, which is also a high abundance order in all of the environmental samples. This suggests that the environment plays some role in establishing the initial bacterial community of the combined ingredients. After two days, the most abundant bacteria present are of the Lactobacillales order, which is expected in the case of successful fermentation. This pattern persists throughout fermentation and jarring.

At the genus level, we find many of the same trends hold (Figure 3B). The three ingredient samples look the most similar at this level, with *Halomonas* common to and prevalent in all three of the samples. The environmental samples continue to show similarity, and the lack of similarity between the environmental samples and the Day 0 fermentation samples persists here as it did at the order level. LAB dominate the other fermentation samples as they did at the order level, with *Leuconostoc* and

Lactobacillus as the dominant genera. This is in line with the results published by Pederson and Albury 1969, which showed *Leuconostoc* and *Lactobacillus* as main players in the sauerkraut fermentation process [12].

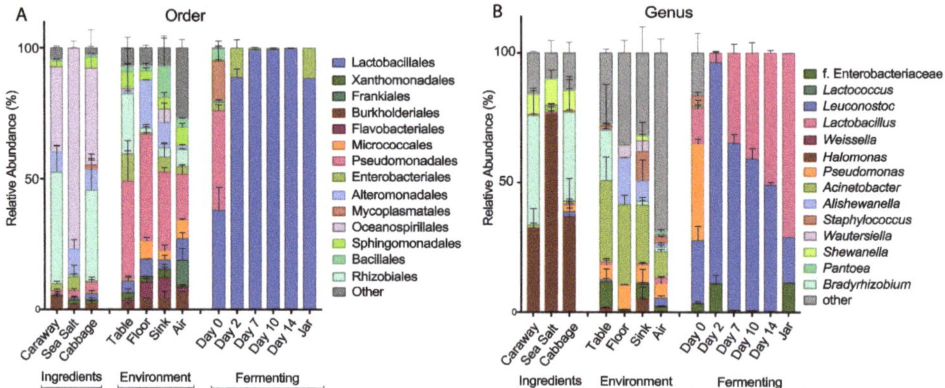

Figure 3. Relative abundance of bacterial taxa in the fermenting sauerkraut, ingredient, and environmental samples at the (**A**) order and (**B**) genus levels. Only the top seven taxa from the fermenting and ingredients/environment sample groupings are shown.

4. Discussion

Overall, our results show that, despite the warmer and more rapid temperature fermentation process used to produce the sauerkraut analyzed here, the bacterial community is in line with that of more traditional, colder fermented cabbage products. Over the first 48 hours of fermentation, the microbial community of sauerkraut experienced a precipitous drop in the number of bacterial taxa present, likely due to the strong selective pressures of high salinity and acidity in the fermentation environment. Over the remainder of the fermentation period, LAB remained the dominant organisms present in the community. Both patterns are indicative of successful fermentation.

Perhaps more surprising were the relationships between the microbial communities of the starting ingredients, the fermentation environment, and the fermenting sauerkraut. The major LAB found in fermenting sauerkraut were present only in extremely low levels in the starting ingredients, which may suggest that only trace amounts of LAB are necessary to initiate fermentation. It is also possible that the abundance of fermentative sauerkraut LAB found around the production facility—especially in the air—might contribute to the inception of the fermentative community, acting as a starter culture. The presence of LAB in the environment may also be a direct result of sauerkraut being fermented within it. These hypotheses require further investigation.

Previous studies have used culture- and sequencing-based methods to elucidate the fermentative microbial community of sauerkraut. Culture-based methods have shown that the major LAB involved in sauerkraut fermentation are *E. faecalis*, *L. mesenteroides*, *L. brevis*, *P. cerevisiae*, and *L. plantarum*; while sequencing-based methods highlight the *Lactobacillus* and *Leuconostoc* species in addition to *Weissella* [11,13]. Our results using 16S rRNA sequencing paralleled these expectations and expanded on previous knowledge, identifying *Leuconostoc*, *Lactobacillus*, and Enterobacteriaceae in addition to a variety of LAB not previously detected, such as *Lactococcus*.

Our results are also in line with the canonical microbial communities of other fermented vegetable foods. Xiong et al. found that *Lactobacillus* and *Leuconostoc* species were the primary bacteria in the fermentation of Chinese sauerkraut, *pàocài* [20]. Numerous studies have shown that the kimchi bacterial community is dominated by *Weisella*, *Lactobacillus*, and *Leuconostoc* species [21–23]. A study of traditional Vietnamese fermented vegetables, such as mustard and beet ferment (*dua muoi*) and fermented eggplant (*cà muối*), found a predominance of *Lactobacillus* species in fermentation [24].

Our results suggest that warmer and more rapid production can yield fermented sauerkraut with a similar microbial community to sauerkraut produced by traditional fermentation methods. This may mean that a quick-fermented process is a viable option for industrial production of fermented cabbage foods. This may be of interest to commercial producers, as it would allow them to speed and scale-up production without sacrificing the integrity of the fermentative bacterial community, which is central to the purported probiotic benefits of sauerkraut and other fermented foods.

While the analyzed communities were roughly similar to previously published sauerkraut data, we cannot yet claim that the products are identical or that the production processes are interchangeable. There are multiple metrics—physical, sensory, and nutritive—that were not investigated as part of this study and could possibly vary between the two types of sauerkraut. We anticipate that diminished appearance, shelf life, taste, and nutritive value of warmer fermented sauerkraut could negatively impact its commercial viability. Therefore, additional studies and measurements of these qualities are required before widespread commercial implementation of this fermentation technique.

Author Contributions: M.A.Z. and P.B. jointly designed this study and collected all samples from the production facility. D.J.C, J.I.W., and M.A.Z. extracted and prepared DNA for sequencing. W.H.S. and D.J.C. performed all analyses. M.A.Z., W.H.S., D.J.C., J.I.W., and P.B. prepared the manuscript. All authors reviewed and approved its final version.

Acknowledgments: Research activities associated with this study were funded in part by the COBRE Center for Computational Biology of Human Disease (NIH P20 GM109035). Sequencing was conducted at the Rhode Island Genomics and Sequencing Center, a Rhode Island NSF EPSCoR research facility supported in part by the NSF ESPCoR Cooperative Agreement #EPS-1004057. M.A.Z was supported by a Karen T. Romer Undergraduate Teaching and Research Fellowship, and D.J.C. was supported by the National Science Foundation Graduate Research Fellowship (Grant No. 1644760).

Conflicts of Interest: The authors declare no conflict of interest.

References

1. Trail, A.C.; Fleming, H.P.; Young, C.T.; McFeeters, R.F. Chemical and Sensory Characterization of Commercial Sauerkraut. *J. Food Qual.* **1996**, *19*, 15–30. [CrossRef]
2. Varzakas, T.; Zakynthinos, G.; Proestos, C.; Radwanska, M. Fermented Vegetables. In *Minimally Processed Refrigerated Fruits and Vegetables*; Yildiz, F., Wiley, R.C., Eds.; Springer: Boston, MA, USA, 2017; pp. 537–584. ISBN 978-1-4939-7016-2.
3. Fleming, H.P.; McFeeters, R.F.; Daeschel, R.F. Fermented and Acidified Vegetables. In *Compendium of Methods for the Microbiological Examination of Foods*; American Public Health Association: Washington, DC, USA, 1992; pp. 929–952. ISBN 978-0-87553-175-5.
4. Stamer, J.R.; Stoyla, B.O.; Dunckel, B.A. Growth Rates and Fermentation Patterns of Lactic Acid Bacteria Associated with the Sauerkraut Fermentation. *J. Milk Food Technol.* **1971**, *34*, 521–525. [CrossRef]
5. Cheigh, H.S.; Park, K.Y. Biochemical, microbiological, and nutritional aspects of kimchi (Korean fermented vegetable products). *Crit. Rev. Food Sci. Nutr.* **1994**, *34*, 175–203. [CrossRef] [PubMed]
6. Settanni, L.; Corsetti, A. Application of bacteriocins in vegetable food biopreservation. *Int. J. Food Microbiol.* **2008**, *121*, 123–138. [CrossRef] [PubMed]
7. Lee, H.; Yoon, H.; Ji, Y.; Kim, H.; Park, H.; Lee, J.; Shin, H.; Holzapfel, W. Functional properties of Lactobacillus strains isolated from kimchi. *Int. J. Food Microbiol.* **2011**, *145*, 155–161. [CrossRef] [PubMed]
8. Turpin, W.; Humblot, C.; Guyot, J.-P. Genetic screening of functional properties of lactic acid bacteria in a fermented pearl millet slurry and in the metagenome of fermented starchy foods. *Appl. Environ. Microbiol.* **2011**, *77*, 8722–8734. [CrossRef] [PubMed]
9. Ljungh, A.; Wadström, T. Lactic acid bacteria as probiotics. *Curr. Issues Intest. Microbiol.* **2006**, *7*, 73–89. [PubMed]
10. Ji, Y.; Kim, H.; Park, H.; Lee, J.; Lee, H.; Shin, H.; Kim, B.; Franz, C.M.A.P.; Holzapfel, W.H. Functionality and safety of lactic bacterial strains from Korean kimchi. *Food Control* **2013**, *31*, 467–473. [CrossRef]

11. Clarke, T.C.; Black, L.I.; Stussman, B.J.; Barnes, P.M.; Nahin, R.L. Trends in the use of complementary health approaches among adults: United States, 2002–2012. *Natl. Health Stat. Rep.* **2015**, *79*, 1–16.
12. Pederson, C.S.; Albury, M.N. *Bulletin: Number 824: The Sauerkraut Fermentation*; Agricultural Experiment Station: New York, NY, USA, 1969.
13. Bell, V.; Ferrão, J.; Fernandes, T. Nutritional Guidelines and Fermented Food Frameworks. *Foods* **2017**, *6*, 65. [CrossRef] [PubMed]
14. Plengvidhya, V.; Breidt, F.; Lu, Z.; Fleming, H.P. DNA fingerprinting of lactic acid bacteria in sauerkraut fermentations. *Appl. Environ. Microbiol.* **2007**, *73*, 7697–7702. [CrossRef] [PubMed]
15. Caporaso, J.G.; Kuczynski, J.; Stombaugh, J.; Bittinger, K.; Bushman, F.D.; Costello, E.K.; Fierer, N.; Peña, A.G.; Goodrich, J.K.; Gordon, J.; et al. QIIME allows analysis of high-throughput community sequencing data. *Nat. Methods* **2010**, *7*, 335–336. [CrossRef] [PubMed]
16. Caporaso, J.G.; Lauber, C.L.; Walters, W.A.; Berg-Lyons, D.; Huntley, J.; Fierer, N.; Owens, S.M.; Betley, J.; Fraser, L.; Bauer, M.; et al. Ultra-high-throughput microbial community analysis on the Illumina HiSeq and MiSeq platforms. *ISME J* **2012**, *6*, 1621–1624. [CrossRef] [PubMed]
17. Walters, W.; Hyde, E.R.; Berg-Lyons, D.; Ackermann, G.; Humphrey, G.; Parada, A.; Gilbert, J.A.; Jansson, J.K.; Caporaso, J.G.; Fuhrman, J.A.; et al. Improved Bacterial 16S rRNA Gene (V4 and V4-5) and Fungal Internal Transcribed Spacer Marker Gene Primers for Microbial Community Surveys. *mSystems* **2016**, *1*, e00009-15. [CrossRef] [PubMed]
18. Thompson, L.R.; Sanders, J.G.; McDonald, D.; Amir, A.; Ladau, J.; Locey, K.J.; Prill, R.J.; Tripathi, A.; Gibbons, S.M.; Ackermann, G.; et al. Earth Microbiome Project Consortium A communal catalogue reveals Earth's multiscale microbial diversity. *Nature* **2017**, *104*, 11436. [CrossRef]
19. Cabral, D.J.; Wurster, J.I.; Flokas, M.E.; Alevizakos, M.; Zabat, M.; Korry, B.J.; Rowan, A.D.; Sano, W.H.; Andreatos, N.; Ducharme, R.B.; et al. The salivary microbiome is consistent between subjects and resistant to impacts of short-term hospitalization. *Sci. Rep.* **2017**, *7*, 11040. [CrossRef] [PubMed]
20. Xiong, T.; Guan, Q.; Song, S.; Hao, M.; Xie, M. Dynamic changes of lactic acid bacteria flora during Chinese sauerkraut fermentation. *Food Control* **2012**, *26*, 178–181. [CrossRef]
21. Lee, M.; Song, J.H.; Jung, M.Y.; Lee, S.H.; Chang, J.Y. Large-scale targeted metagenomics analysis of bacterial ecological changes in 88 kimchi samples during fermentation. *Food Microbiol.* **2017**, *66*, 173–183. [CrossRef] [PubMed]
22. Kim, M.; Chun, J. Bacterial community structure in kimchi, a Korean fermented vegetable food, as revealed by 16S rRNA gene analysis. *Int. J. Food Microbiol.* **2005**, *103*, 91–96. [CrossRef] [PubMed]
23. Zabat, M.A.; Sano, W.H.; Cabral, D.J.; Wurster, J.I.; Belenky, P. The impact of vegan production on the kimchi microbiome. *Food Microbiol.* **2018**, *74*, 171–178. [CrossRef] [PubMed]
24. Nguyen, D.T.L.; Van Hoorde, K.; Cnockaert, M.; De Brandt, E.; Aerts, M.; Binh Thanh, L.; Vandamme, P. A description of the lactic acid bacteria microbiota associated with the production of traditional fermented vegetables in Vietnam. *Int. J. Food Microbiol.* **2013**, *163*, 19–27. [CrossRef] [PubMed]

 *foods*

Article

Lactobacillus plantarum with Broad Antifungal Activity as a Protective Starter Culture for Bread Production

Pasquale Russo [1,2,*], **Clara Fares** [3], **Angela Longo** [1], **Giuseppe Spano** [1] and **Vittorio Capozzi** [1]

[1]   Department of Science of Agriculture, Food and Environment, University of Foggia, Via Napoli 25, 71122 Foggia, Italy; angela.longo@unifg.it (A.L.); giuseppe.spano@unifg.it (G.S.); vittorio.capozzi@unifg.it (V.C.)

[2]   Promis Biotech Via Napoli 25, 71122 Foggia, Italy

[3]   Council for Agricultural Research and Economics—Research Centre for Cereal and Industrial Crops (CREA-CI), S.S.673 km 25.200, 71122 Foggia, Italy; clara.fares@crea.gov.it

*   Correspondence: pasquale.russo@unifg.it; Tel.: +39-0881-589-303

Received: 13 November 2017; Accepted: 4 December 2017; Published: 11 December 2017

Abstract: Bread is a staple food consumed worldwide on a daily basis. Fungal contamination of bread is a critical concern for producers since it is related to important economic losses and safety hazards due to the negative impact of sensorial quality and to the potential occurrence of mycotoxins. In this work, *Lactobacillus plantarum* UFG 121, a strain with characterized broad antifungal activity, was analyzed as a potential protective culture for bread production. Six different molds belonging to *Aspergillus* spp., *Penicillium* spp., and *Fusarium culmorum* were used to artificially contaminate bread produced with two experimental modes: (i) inoculation of the dough with a commercial *Saccharomyces cerevisiae* strain (control) and (ii) co-inoculation of the dough with the commercial *S. cerevisiae* strain and with *L. plantarum* UFG 121. *L. plantarum* strain completely inhibited the growth of *F. culmorum* after one week of storage. The lactic acid bacterium modulated the mold growth in samples contaminated with *Aspergillus flavus*, *Penicillium chrysogenum*, and *Penicillium expansum*, while no antagonistic effect was found against *Aspergillus niger* and *Penicillium roqueforti*. These results indicate the potential of *L. plantarum* UFG 121 as a biocontrol agent in bread production and suggest a species- or strain-depending sensitivity of the molds to the same microbial-based control strategy.

Keywords: antifungal; bioprotection; bread; *Lactobacillus plantarum*; phenyllactic acid; *Aspergillus*; *Penicillium*; *Fusarium*

1. Introduction

Bread, obtained by baking a fermented dough of cereals flour, water, and other ingredients, is ancient and, due to its nutritional properties and low price, is a staple of many diets and an essential contributor of energy and nutritional intake in both developed and developing countries [1,2]. Microbial alteration of bread is a critical concern for bakeries, and it is mainly attributable to the development of spoilage molds. Apart from significant economic losses due to the negative impact on sensory properties, the occurrence of filamentous fungi poses a safety hazard for human health due to the potential ability of some fungal strains to produce mycotoxins [3,4]. Moreover, fungal spoilage control is critical for the extension of the shelf life of bakery goods, especially from an industrial perspective [5]. Traditionally, the shelf life of bread has been extended by the addition of chemical preservatives such as ethanol and weak organic acids, mainly propionic, sorbic, benzoic, and acetic acid and their salts [6]. As an alternative, physical methods such as microwave and infrared radiation and innovative packaging technologies have been exploited to reduce fungal developments in bakery products [7]. However, a strong societal demand, supported by public authorities, has urged more eco-friendly approaches mainly relying on the use of essential oils and antagonistic microorganisms as preservation tools [8]. In this context, lactic acid bacteria (LAB) have the greatest appeal as biocontrol agents due to their status of food-grade microorganisms [9,10]. The antifungal ability of some LAB strains is owed to the production of secondary metabolites, mainly including lactic acid and other organic acids, phenolic compounds, carbon dioxide, ethanol, hydrogen peroxide, fatty acids, acetoin, diacetyl, and cyclic dipeptides [11–13]. Moreover, synergistic interactions among different bioactive molecules could substantially increase the overall LAB antimicrobial activity [14]. Sourdough is a valuable and comprehensive source of antifungal and mycotoxin-controlling compounds synthesized by LAB during fermentation [15–17]. In the last few years, several LAB strains belonging to the species *Lactobacillus amylovorus*, *Lactobacillus reuteri*, *Lactobacillus brevis*, *Lactobacillus plantarum*, *Lactobacillus rossiae*, and *Lactobacillus paralimentarius* have been proposed as starter protective cultures to enhance the shelf life of bread [18–22]. Antifungal properties against bakery product spoilage molds have also been shown by *Propionibacterium* cultures [23]. Moreover, the ability of antagonistic yeasts to control fungal contamination in bread has been investigated in *Meyerozyma guilliermondii* and *Wickerhamomyces anomalus* [24,25]. In particular, *Penicillium roqueforti* was delayed until 14 days of storage in bread produced with a combination of these antifungal yeasts and of a specific *L. plantarum* strain [25]. Furthermore, proteinaceous compounds from different food matrices or legumes flour hydrolysates could be used as ingredients in the bakery industry to enhance the antifungal properties of sourdoughs [26,27].

In this work, a strain of *L. plantarum* previously characterized for its antifungal potential [28] was investigated for its ability to control the growth of six different species of filamentous fungi belonging to three different genera in artificially contaminated bread after one week of storage.

2. Materials and Methods

2.1. Microbial Strains and Growth Conditions

Lactobacillus plantarum UFG 121 was available at the Laboratory of Industrial Microbiology of the University of Foggia (Foggia, Italy) and routinely grown in de Man Rogosa Sharpe (MRS) broth (Oxoid, Basingstoke, UK) at 30 °C for 24 h.

Six filamentous fungi, namely *Penicillium roqueforti* CECT 20508, *Penicillium expansum* CECT 2278, *Penicillium chrysogenum* CECT 2669, *Aspergillus niger* CECT 2805, *Aspergillus flavus* CECT 20802, and *Fusarium culmorum* CECT 2148 were provided by the Spanish Type Culture Collection (CECT, Paterna, Spain). Fungal cultures were plated on malt extract agar (Oxoid) and incubated at 24 °C for 5 days.

The commercial fresh yeast *Saccharomyces cerevisiae* "Lievital" (Lessafre, Marcq-en-Baroeul, France) was resuspended in sterile saline solution, streaked on plates of yeast extract, peptone, dextrose (YPD, Oxoid), and incubated at 30 °C for 48 h. Then, a single colony was resuspended in YPD broth and incubated at 30 °C.

2.2. Dough and Sourdough Preparation

Cells at exponential phase of *S. cerevisiae* and *L. plantarum* UFG 121 were recovered by centrifugation (5000× *g*, 5 min), washed twice with sterile saline solution, and resuspended in sterile water. For sourdough preparation, microbial starters were inoculated at a concentration of about 10^8 and 10^6 cfu g^{-1} for *L. plantarum* and yeast respectively, in a mixture of wheat flour type "0" Manitoba (initial moisture of 11.9% *w/w*, supplied by Lo Conte, IPAFOOD, Ariano Irpino, Italy) and water (37.5% *w/w*) containing sucrose (6%), NaCl (3% *w/w*), and animal fats (3% *w/w*). A dough control sample was inoculated only with the commercial yeast at a final concentration of about 10^6 cfu g^{-1}. The fermentation was carried out at 30 °C for 18 h.

2.3. Bread Production

Bread was obtained in a pilot plant according to the methods ACC10-10B29 modified by Capozzi et al. [29]. Briefly, the sourdough was subject to kneading for 10 min, followed by fermentation for 90 min at room temperature. The sourdough was then aliquoted in samples of about 250 g each and fermented for another 90 min. Afterwards, samples were put into apposite shapes to attain specific dimensions (11 cm × 25 cm × 7 cm) and submitted to a final fermentation step of 90 min at room temperature. Bread was obtained after the dough was baked in an oven at 220 °C for 40 min. The bread was cut in half in order to repeat each experimental condition in duplicate.

2.4. Artificial Contamination of Bread

A preparation of fungal spores was obtained by brushing with a sterile swab the plate surface of each five-day-grown mold. Spores were resuspended in sterile distilled water and concentrated at 8×10^4 spores mL^{-1}. Artificially contaminated samples were obtained by spraying 15 mL of the spore solution on the surface of the bread. Samples were immediately packaged using polyethylene terephthalate bags and stored for 7 days at room temperature. At this time, the in vivo antagonistic activity against each tested mold was qualitatively determined by comparing the area contaminated by the spoilage fungi in bread fermented with the starter yeast or co-fermented with *L. plantarum* UFG 121. Results were expressed according to the following scale: no/low (−), moderate (+), high (+ +), and very high (+ + +) inhibition activity, if the area covered by each filamentous fungi in bread co-fermented with UFG 121 strain was reduced in the ranges 0–25%, 25–50%, 50–75%, and 75–100%, respectively.

2.5. Sensorial Quality

The sensorial analysis was carried out at our laboratory by a panel of six panelists before the artificial contamination of the samples with the spoilage molds. Panelists were previously trained in order to recognize and score the analyzed quality descriptors. Samples were coded with a random 3-digit number in order to minimize subjectivity. The sensorial attributes evaluated were overall appearance, aromatic profile, off odor, color, softness, and texture. Every attribute was scored on a 1–5 hedonic scale, where 1 = atypical, undesirable, and 5 = typical, desired. Sensorial trials were performed on six samples for each treatment.

2.6. Statistical Analysis

The quality descriptors analyzed in the sensorial trials were subjected to one-way analysis of variance (ANOVA). Pairwise comparison of treatment means was achieved by Tukey's procedure, with a significance level of $p \leq 0.05$, using the statistical software Past 3.0 (University of Oslo, Oslo, Norway).

3. Results

3.1. Production of Bread Co-Fermented with a Protective L. plantarum Strain

Bread samples were produced by fermentation of the dough with a commercial yeast commonly used in breadmaking, or by its co-inoculation with *L. plantarum* UFG 121, a strain with a characterized antifungal activity [28]. A preliminary qualitative analysis was performed during the breadmaking process in order to detect if the employment of the protective strain *L. plantarum* UFG 121 as a starter culture could affect the bread production from a technological and/or sensorial point of view. As reported in Figure 1, some differences were observed among samples inoculated with the commercial yeast starter or co-inoculated with UFG 121 strain. In particular, after 18 h of the fermentation step, the dough was apparently softer when fermented with *S. cerevisiae* compared with the co-inoculation approach using the protective *L. plantarum* strain (Figure 1A). In contrast, the sourdough developed a more complex aromatic profile when co-fermented with UFG 121 strain. In agreement with this finding, after kneading the dough fermented using the yeast starter, the dough was more compact and elastic and therefore easier to break and to produce the desired shape. Co-fermentation with yeast and *L. plantarum* resulted in a sourdough with an inhomogeneous texture, which could be disadvantageous for subsequent processing (Figure 1B). However, no important differences were detected after the final fermentation, since both samples were well compacted and leavened with a smooth surface (Figure 1C). After all samples were baked, similar features in terms of color, softness, and texture were apparent. However, samples co-inoculated with *L. plantarum* UFG 121 presented more pronounced alveolation (Figures 1D and 2). Moreover, sensorial analysis indicated the absence of off odors in both breads, while the aromatic profile of the bread inoculated with *L. plantarum* UFG 121 scored higher than the control bread. In general, the overall appearance of the bread obtained via UFG 121 co-fermentation was perceived as slightly better by the panelists.

Figure 1. Dough fermented by *S. cerevisiae* "Lievital" (left pictures) or co-fermented with *L. plantarum* UFG 121 (right pictures) after (**A**) first fermentation step for 18 h at 30 °C; (**B**) kneading and shaping; (**C**) the final fermentation step for 90 min at room temperature; and (**D**) baking and cutting.

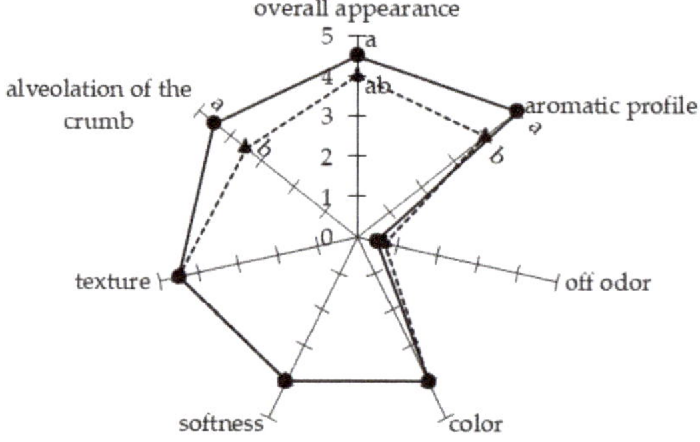

Figure 2. Sensory attributes of bread obtained by fermentation with *S. cerevisiae* "Lievital" (triangle, dashed line) or co-fermented with *L. plantarum* UFG 121 (circle, continuous line). Reported values are means of six independent replicates, and they are expressed by using a hedonic scale from 1 to 5 (1 = atypical, undesirable; 5 = typical, desired). Means with different letters are significantly different according to Tukey's test ($p \leq 0.05$).

3.2. Analysis of the Protective Potential of L. plantarum UFG 121 in Artificially Contaminated Bread

With the aim to evaluate the potential of *L. plantarum* UFG 121 as a culture protective against typical molds on bakery products, bread samples were artificially contaminated by spraying a concentrated spore suspension onto the bread slices. After one week of storage, the in vivo antagonistic activity against each tested mold was qualitatively determined by comparing the area covered by the spoilage fungi in the two tested experimental modes: (1) bread fermented with the commercial *S. cerevisiae* and (2) co-fermented with *L. plantarum* UFG 121 (Figure 3). As shown in Figure 3, after one week of storage the surface of the control bread artificially contaminated were always wholly covered by the molds. In contrast, different scenarios were observed when bread samples were co-fermented with *L. plantarum* UFG 121. In particular, no inhibition was found in bread samples artificially inoculated with *A. niger* and *P. roqueforti* that appeared completely contaminated by both the molds (Figure 3A,B). A moderate in vivo antagonistic activity was detected against *P. chrysogenum* and *P. expansum* whose growth was limited in samples obtained with UFG 121 strain (Figure 3C,D). In contrast, a higher protective effect was observed in samples contaminated by *A. flavus* in which only approximately 20% of the bread surface was covered by the spoilage (Figure 3E). Interestingly, no development of *F. culmorum* was observed, suggesting that the employment of *L. plantarum* UFG 121 during bread fermentation was a successful strategy to thoroughly inhibit *F. culmorum* growth (Figure 3F).

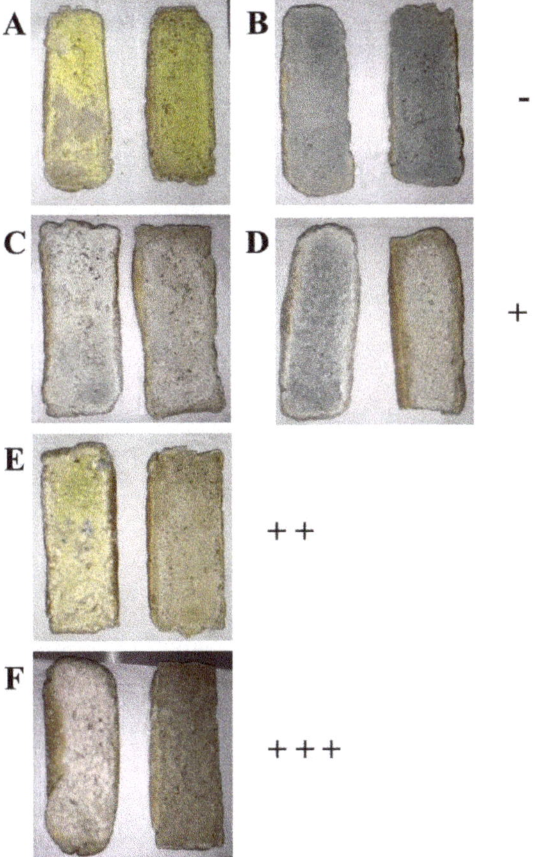

Figure 3. Bread obtained by fermentation with *S. cerevisiae* "Lievital" (left pictures) or co-fermented with *L. plantarum* UFG 121 (right pictures) after one week of storage at room temperature and artificially contaminated with *A. niger* CECT 2805 (**A**); *P. roqueforti* CECT 20508 (**B**); *P. chrysogenum* CECT 2669 (**C**); *P. expansum* CECT 2278 (**D**); *A. flavus* CECT 20802 (**E**); *F. culmorum* CECT 2148 (**F**). Antifungal activity was expressed as no/low (−), moderate (+), high (+ +), and very high (+ + +), when the contaminated area was reduced in the ranges 0–25%, 25–50%, 50–75%, and 75–100%, respectively.

4. Discussion

In the present study, six different molds belonging to *Aspergillus* spp., *Penicillium* spp., and *Fusarium culmorum* were used to artificially contaminate bread produced by fermentation of the dough with a commercial *S. cerevisiae*, or by its co-inoculation with *L. plantarum* UFG 121. The fungal strains were selected because they are representative molds of bread spoilage [30] and they have an ability to produce mycotoxins. In particular, *A. flavus* CECT 20802 produced aflatoxins B1, B2, M1, M2; *F. culmorum* CECT 2148 fusarine C; *P. roqueforti* CECT 20508 was able to synthesize PR toxin; while *P. expansum* CECT 2278 was responsible for the production of patulin and citrinin. In the present study, *L. plantarum* UFG 121, previously characterized for its broad antifungal activity against these mold strains [28], was investigated as a protective culture for bread production.

The employment of LAB strains with antagonistic activity has been widely proposed as an innovative green strategy to fight spoilage filamentous fungi in order to enhance the shelf life of bakery products [24,31–33]. In general, the antifungal effect has been attributed to the production of some secondary metabolites during the fermentation of sourdough [11]. Therefore, according to previous studies [24,34], in this work, a fermentation step of 18 h was performed in order to allow the UFG 121 strain to enrich the sourdoughs with bioactive antifungal compounds. However, differences between the control samples and the bread obtained by co-fermentation with *L. plantarum* UFG 121 were qualitatively detected as the bread was made that might have affected the production process. Nonetheless, as previous reported by Coda et al. [25], co-fermentation of the dough with yeast and selected LAB resulted in bread showing good chemical and textural features, including elasticity, color, and alveolation. Moreover, an improvement in terms of complexity of the aromatic profile was noted in the bread co-fermented with UFG 121. In agreement with this finding, Makhoul et al. [35], analyzing pro-technological microbe/matrix interactions during food fermentation, reported a greater impact of the microbial fraction on the volatile organic compounds of bread. However, the present work is only a preliminary study that should be further complemented by the analytical determination of the main physico-chemical parameters as well as the impact of a protective LAB culture on the organoleptic profile of the bread [36,37].

In our previous study, we found that the preservative potential of *L. plantarum* UFG 121 was mainly due to the production of lactic acid and phenyllactic acid [28]. Phenyllactic acid production by *L. plantarum* has been linked to the antifungal activity against fungal strains isolated from bakery products belonging to species of *Aspergillus*, *Penicillium*, and *Fusarium* [38]. Similarly, an increase in the shelf life of bread obtained by fermentation with *L. plantarum* CRL 778 and artificially contaminated with *Penicillium* spp. has been found to be related to the synthesis of acetic and phenyllactic acid as well as lactic acid [20]. Moreover, organic acids including phenyllactic acid from a strain of *L. amylovorus* have been found to be responsible for a higher shelf life in gluten-free breads [21]. Lactic acid, phenyllactic acid, and two cyclic dipeptides found in sourdoughs fermented by *L. plantarum* FST 1.7 have been identified as the main antifungal compounds able to retard the growth of *F. culmorum* and *Fusarium graminearum* on bread [18].

In a similar way, in the present study, *F. culmorum* CECT 2148 was completely inhibited in bread fermented by *L. plantarum* UFG 121. The in vitro assays performed suggest that CECT 2148 was the most sensitive mold when exposed to UFG 121 cell-free supernatant [28]. Moreover, this result has been confirmed in situ, since fermentation by UFG 121 and artificial contamination with CECT 2148 (after thermal stabilization) increased the shelf life of an oat-based formulation from less than one week to the second week of cold storage, indicating that a strong bioprotection could be provide by antifungal compounds produced in a 16 h fermentation step [28]. In contrast, fermentation with *L. plantarum* UFG 121 had no effect in countering the growth of *P. roqueforti* CECT 20508 and *A. niger* CECT 2805, but reduced bread contamination by the remaining tested molds to different extents. Interestingly, these results were only partially predicted by the in vitro assays [28], suggesting that interactions with the commercial yeast, process parameters, and/or the food matrix might modulate the antagonistic activity of selected LAB against fungal strains. Therefore, further investigations are required to establish the effectiveness of the antifungal compounds, their synergistic interactions, and the complex microbial and physico-chemical relationships occurring in the food environment.

5. Conclusions

In recent years, several studies have aimed to enhance the shelf life and safety of bakery products by analyzing the antifungal potential of protective microbial starter cultures and the corresponding sourdough. However, most of these works have been performed using only one or a few mold strains as fungal indicators. In this study, although with a preliminary experimental plan, we analyzed the in situ bioprotection effectiveness of a *L. plantarum* strain against a representative panel of six different mold species belonging to three different genera generally recognized as being mainly responsible for the natural contamination of bakery foods. Our results indicated in these molds a species- or strain-dependent sensitivity to the same microbial-based control strategy, suggesting that a new generation of mixed starter protective cultures, active against different fungal species, might be better able to globally extend the shelf life of baked goods.

Acknowledgments: Vittorio Capozzi was supported by Fondo di Sviluppo e Coesione 2007-2013—APQ Ricerca Regione Puglia "Programma regionale a sostegno della specializzazione intelligente e della sostenibilità sociale ed ambientale—FutureInResearch". Pasquale Russo was supported by a grant of the Apulian Region in the framework of "Perform Tech (Puglia Emerging Food Technology)" project (practice code LPIJ9P2).

Author Contributions: P.R., V.C., and G.S. conceived and designed the experiments; A.L. performed the experiments; P.R. and V.C. analyzed the data; C.F. contributed to the bread production in the pilot plant; P.R. wrote the paper; G.S. and V.C. critically revised the manuscript.

Conflicts of Interest: The authors declare no conflict of interest.

References

1. Guyot, J.P. Cereal-based fermented foods in developing countries: Ancient foods for modern research. *Int. J. Food Sci. Technol.* **2012**, *47*, 1109–1114. [CrossRef]
2. Silow, C.; Axel, C.; Zannini, E.; Arendt, E.K. Current status of salt reduction in bread and bakery products—A review. *J. Cereal Sci.* **2016**, *72*, 135–145. [CrossRef]
3. Vaclavikova, M.; Malachova, A.; Veprikova, Z.; Dzuman, Z.; Zachariasova, M.; Hajslova, J. 'Emerging' mycotoxins in cereals processing chains: Changes of enniatins during beer and bread making. *Food Chem.* **2013**, *136*, 750–757. [CrossRef] [PubMed]
4. Lee, H.J.; Ryu, D. Worldwide occurrence of mycotoxins in cereals and cereal-derived food products: Public health perspectives of their co-occurrence. *J. Agric. Food Chem.* **2017**, *65*, 7034–7051. [CrossRef] [PubMed]
5. Almeida, E.L.; Steel, C.J.; Chang, Y.K. Par-baked bread technology: Formulation and process studies to improve quality. *Crit. Rev. Food Sci. Nutr.* **2016**, *56*, 70–81. [CrossRef] [PubMed]
6. Samapundo, S.; Devlieghere, F.; Vroman, A.; Eeckhout, M. Antifungal properties of fermentates and their potential to replace sorbate and propionate in pound cake. *Int. J. Food Microbiol.* **2016**, *237*, 157–163. [CrossRef] [PubMed]
7. Van Long, N.N.; Joly, C.; Dantigny, P. Active packaging with antifungal activities. *Int. J. Food Microbiol.* **2016**, *220*, 73–90. [CrossRef] [PubMed]
8. Salas, M.L.; Mounier, J.; Valence, F.; Coton, M.; Thierry, A.; Coton, E. Antifungal Microbial agents for food biopreservation—A review. *Microorganisms* **2017**, *5*, 3.
9. European Food Safety Authority. Introduction of a qualified presumption of safety (QPS) approach for assessment of selected microorganisms referred to EFSA. *EFSA J.* **2007**, *587*, 1–16.
10. Varsha, K.K.; Nampoothiri, K.M. Appraisal of lactic acid bacteria as protective cultures. *Food Control* **2016**, *69*, 61–64. [CrossRef]
11. Messens, W.; de Vuyst, L. Inhibitory substances produced by *Lactobacilli* isolated from sourdoughs—A review. *Int. J. Food Microbiol.* **2002**, *72*, 31–43. [CrossRef]
12. Crowley, S.; Mahony, J.; Van Sinderen, D. Current perspectives on antifungal lactic acid bacteria as natural bio-preservatives. *Trends Food Sci. Technol.* **2013**, *33*, 93–109. [CrossRef]

13. Le Lay, C.; Coton, E.; Le Blay, G.; Chobert, J.M.; Haertlé, T.; Choiset, Y.; Van Long, N.N.; Meslet-Cladière, L.; Mounier, J. Identification and quantification of antifungal compounds produced by lactic acid bacteria and propionibacteria. *Int. J. Food Microbiol.* **2016**, *239*, 79–85. [CrossRef] [PubMed]

14. Cortés-Zavaleta, O.; López-Malo, A.; Hernández-Mendoza, A.; García, H.S. Antifungal activity of lactobacilli and its relationship with 3-phenyllactic acid production. *Int. J. Food Microbiol.* **2014**, *173*, 30–35. [CrossRef] [PubMed]

15. Lavermicocca, P.; Valerio, F.; Evidente, A.; Lazzaroni, S.; Corsetti, A.; Gobbetti, M. Purification and characterization of novel antifungal compounds from the sourdough *Lactobacillus plantarum* strain 21B. *Appl. Environ. Microbiol.* **2000**, *66*, 4084–4090. [CrossRef] [PubMed]

16. Hassan, Y.I.; Zhou, T.; Bullerman, L.B. Sourdough lactic acid bacteria as antifungal and mycotoxin-controlling agents. *Food Sci. Technol. Int.* **2016**, *22*, 79–90. [CrossRef]

17. Saladino, F.; Luz, C.; Manyes, L.; Fernández-Franzón, M.; Meca, G. In vitro antifungal activity of lactic acid bacteria against mycotoxigenic fungi and their application in loaf bread shelf life improvement. *Food Control* **2016**, *67*, 273–277. [CrossRef]

18. Dal Bello, F.; Clarke, C.I.; Ryan, L.A.M.; Ulmer, H.; Schober, T.J.; Ström, K.; Sjögren, J.; van Sinderen, D.; Schnürer, J.; Arendt, E.K. Improvement of the quality and shelf life of wheat bread by fermentation with the antifungal strain *Lactobacillus plantarum* FST 1.7. *J. Cereal Sci.* **2007**, *45*, 309–318. [CrossRef]

19. Rizzello, C.G.; Cassone, A.; Coda, R.; Gobbetti, M. Antifungal activity of sourdough fermented wheat germ used as an ingredient for bread making. *Food Chem.* **2011**, *127*, 952–959. [CrossRef] [PubMed]

20. Gerez, C.L.; Torino, M.I.; Rollán, G.; Font de Valdez, G. Prevention of bread mould spoilage by using lactic acid bacteria with antifungal properties. *Food Control* **2009**, *20*, 144–148. [CrossRef]

21. Axel, C.; Röcker, B.; Brosnan, B.; Zannini, E.; Furey, A.; Coffey, A.; Arendt, E.K. Application of *Lactobacillus amylovorus* DSM19280 in gluten-free sourdough bread to improve the microbial shelf life. *Food Microbiol.* **2015**, *47*, 36–44. [CrossRef] [PubMed]

22. Axel, C.; Brosnan, B.; Zannini, E.; Peyer, L.C.; Furey, A.; Coffey, A.; Arendt, E.K. Antifungal activities of three different *Lactobacillus* species and their production of antifungal carboxylic acids in wheat sourdough. *Appl. Microbiol. Biotechnol.* **2016**, *100*, 1701–1711. [CrossRef] [PubMed]

23. Le Lay, C.; Mounier, J.; Vasseur, V.; Weill, A.; Le Blay, G.; Barbier, G.; Coton, E. In vitro and in situ screening of lactic acid bacteria and propionibacteria antifungal activities against bakery product spoilage molds. *Food Control* **2016**, *60*, 247–255. [CrossRef]

24. Coda, R.; Cassone, A.; Rizzello, C.G.; Nionelli, L.; Cardinali, G.; Gobbetti, M. Antifungal activity of *Wickerhamomyces anomalus* and *Lactobacillus plantarum* during sourdough fermentation: Identification of novel compounds and long-term effect during storage of wheat bread. *Appl. Environ. Microbiol.* **2011**, *77*, 3484–3492. [CrossRef] [PubMed]

25. Coda, R.; Rizzello, C.G.; Di Cagno, R.; Trani, A.; Cardinali, G.; Gobbetti, M. Antifungal activity of *Meyerozyma guilliermondii*: Identification of active compounds synthesized during dough fermentation and their effect on long-term storage of wheat bread. *Food Microbiol.* **2013**, *33*, 243–251. [CrossRef] [PubMed]

26. Coda, R.; Rizzello, C.G.; Nigro, F.; De Angelis, M.; Arnault, P.; Gobbetti, M. Long-term fungal inhibitory activity of water-soluble extracts of *Phaseolus vulgaris* cv. Pinto and sourdough lactic acid bacteria during bread storage. *Appl. Environ. Microbiol.* **2008**, *74*, 7391–7398. [CrossRef] [PubMed]

27. Rizzello, C.G.; Lavecchia, A.; Gramaglia, V.; Gobbetti, M. Long-term fungal inhibition by *Pisum sativum* flour hydrolysate during storage of wheat flour bread. *Appl. Environ. Microbiol.* **2015**, *81*, 4195–4206. [CrossRef] [PubMed]

28. Russo, P.; Arena, M.P.; Fiocco, D.; Capozzi, V.; Drider, D.; Spano, G. *Lactobacillus plantarum* with broad antifungal activity: A promising approach to increase safety and shelf-life of cereal-based products. *Int. J. Food Microbiol.* **2017**, *247*, 48–54. [CrossRef] [PubMed]

29. Capozzi, V.; Menga, V.; Digesu, A.M.; De Vita, P.; van Sinderen, D.; Cattivelli, L.; Fares, C.; Spano, G. Biotechnological production of vitamin B2-enriched bread and pasta. *J. Agric. Food Chem.* **2011**, *59*, 8013–8020. [CrossRef] [PubMed]

30. Axel, C.; Zannini, E.; Arendt, E.K. Mold spoilage of bread and its biopreservation: A review of current strategies for bread shelf life extension. *Crit. Rev. Food Sci. Nutr.* **2017**, *57*, 3528–3542. [CrossRef] [PubMed]

31. Gerez, C.L.; Torino, M.I.; Obregozo, M.D.; Font de Valdez, G. A ready-to-use antifungal starter culture improves the shelf life of packaged bread. *J. Food Prot.* **2010**, *73*, 758–762. [CrossRef] [PubMed]

32. Garofalo, C.; Zannini, E.; Aquilanti, L.; Silvestri, G.; Fierro, O.; Picariello, G.; Clementi, F. Selection of sourdough lactobacilli with antifungal activity for use as biopreservatives in bakery products. *J. Agric. Food Chem.* **2012**, *60*, 7719–7728. [CrossRef] [PubMed]
33. Axel, C.; Brosnan, B.; Zannini, E.; Furey, A.; Coffey, A.; Arendt, E.K. Antifungal sourdough lactic acid bacteria as biopreservation tool in quinoa and rice bread. *Int. J. Food Microbiol.* **2016**, *239*, 86–94. [CrossRef] [PubMed]
34. Russo, P.; Capozzi, V.; Arena, M.P.; Spadaccino, G.; Dueñas, M.T.; López, P.; Fiocco, D.; Spano, G. Riboflavin-overproducing strains of *Lactobacillus fermentum* for riboflavin-enriched bread. *Appl. Microbiol. Biotechnol.* **2014**, *98*, 3691–3700. [CrossRef] [PubMed]
35. Makhoul, S.; Romano, A.; Capozzi, V.; Spano, G.; Aprea, E.; Cappellin, L.; Benozzi, E.; Scampicchio, M.; Märk, T.D.; Gasperi, F.; et al. Volatile compound production during the bread-making process: Effect of flour, yeast and their interaction. *Food Bioprocess Technol.* **2015**, *8*, 1925–1937. [CrossRef]
36. Capozzi, V.; Makhoul, S.; Aprea, E.; Romano, A.; Cappellin, L.; Sanchez Jimena, A.; Spano, G.; Gasperi, F.; Scampicchio, M.; Biasioli, F. PTR-MS characterization of VOCs associated with commercial aromatic bakery yeasts of wine and beer origin. *Molecules* **2016**, *21*, 483. [CrossRef] [PubMed]
37. Ficco, D.B.M.; Saia, S.; Beleggia, R.; Fragasso, M.; Giovanniello, V.; De Vita, P. Milling overrides cultivar, leavening agent and baking mode on chemical and rheological traits and sensory perception of durum wheat breads. *Sci. Rep.* **2017**, *7*, 13632. [CrossRef] [PubMed]
38. Lavermicocca, P.; Valerio, F.; Visconti, A. Antifungal activity of phenyllactic acid against molds isolated from bakery products. *Appl. Environ. Microbiol.* **2003**, *69*, 634–640. [CrossRef] [PubMed]

Article

Nutritional Guidelines and Fermented Food Frameworks

Victoria Bell [1], Jorge Ferrão [2] and Tito Fernandes [3],*

[1] Faculty of Pharmacy, Coimbra University, Pólo das Ciências da Saúde, 3000-548 Coimbra, Portugal; victoriabell1103@gmail.com

[2] The Vice-Chancellor's Office, Universidade Pedagógica, Rua João Carlos Raposo Beirão 135, Maputo, Moçambique; ljferrao@icloud.com

[3] Associação para o Desenvolvimento das Ciências Veterinárias (ACIVET), Faculty of Veterinary Medicine, Lisbon University, 1300-477 Lisboa, Portugal

* Correspondence: profcattitofernandes@gmail.com; Tel.: +351-919-927930

Received: 26 June 2017; Accepted: 2 August 2017; Published: 7 August 2017

Abstract: This review examines different nutritional guidelines, some case studies, and provides insights and discrepancies, in the regulatory framework of Food Safety Management of some of the world's economies. There are thousands of fermented foods and beverages, although the intention was not to review them but check their traditional and cultural value, and if they are still lacking to be classed as a category on different national food guides. For understanding the inconsistencies in claims of concerning fermented foods among various regulatory systems, each legal system should be considered unique. Fermented foods and beverages have long been a part of the human diet, and with further supplementation of probiotic microbes, in some cases, they offer nutritional and health attributes worthy of recommendation of regular consumption. Despite the impact of fermented foods and beverages on gastro-intestinal wellbeing and diseases, their many health benefits or recommended consumption has not been widely translated to global inclusion in world food guidelines. In general, the approach of the legal systems is broadly consistent and their structures may be presented under different formats. African traditional fermented products are briefly mentioned enhancing some recorded adverse effects. Knowing the general benefits of traditional and supplemented fermented foods, they should be a daily item on most national food guides.

Keywords: fermented foods; nutritional guidelines; legislation; national food guides

1. Introduction

Human nutrition begins with milk. Fermented milk products have been recognized as healthy foods since ancient times. Fermentation processes and products are believed to have been developed 9000 years ago in order to preserve food for times of deficiency, improve flavor, and reduce poisonous effects. Recommendations for the consumption of certain nutritious foods date back to the Hippocratic Corpus of Ancient Greece [1]. Thousands of different fermented foods and beverages are still unknown outside the native area in which they have been produced for centuries, many going back even before recorded history [2]. Fermented foods and beverages pass through a process of lacto fermentation in which natural bacteria or yeasts feed on the sugar and starch in the food creating lactic acid.

The list of fermented products is extremely vast and the diversity derives from the heterogeneity of traditions found in the world, the cultural preference, different geographical areas where they are produced and the staple and/or by-products used for fermentation. The most popular involve beverages such as wine, beer, cider and foods such as yoghurt, cheese, soya, beans, fish, meat, cabbages, among others. In many instances, it is highly likely that the methods of production were unknown and came about by chance, and were passed down by cultural traditional values to subsequent generations [3].

Modern food is submitted to many processing methods such as pasteurization, affecting its nutritional value by reducing vitamins, fiber, minerals, essential fatty acids and amino acids. Food security can be enhanced in poor rural areas with fermented products, generating income in a small-scale family farm in developing countries [4]. The importance of fermentation is reflected by the amount and variety of foods and beverages traded not only for the benefits on nutrition and health-promoting effects but also for preservation, safety, and their peculiar appreciated sensory attributes [5].

The exploration of the microbial communities and enzymes of fermented products has been extensively reviewed [6]. At the genus level, *Lactobacillus* is usually the most abundant genus, followed by *Lactococcus*, *Enterococcus*, *Vibrio*, *Weissella*, *Pediococcus*, *Enterobacter*, *Salinivibrio*, *Acinetobacter*, *Macrococcus*, *Kluyvera* and *Clostridium*.

A better knowledge of microorganisms and fermentation at a molecular level is still required to support and develop the production of sustainable fermented food with high nutritional characteristics. A metagenomic approach has enabled identification of novel microbiome profiles and exploration of microbial compositions in a range of traditional fermented foods while bypassing the need for cultivation, allowing the identification of a vast array of microorganisms never previously isolated in culture [7].

2. Sustainable Development and Fermented Food Hygiene in Africa

Food preservation increases the range of raw materials and by-products that can be used to produce edible food products and remove anti-nutritional factors, rendering food safe to eat by humans and animals. Fermentation is a cheap way of preserving perishable raw materials, accessible to even the most marginalized people. Utilizing small-scale fermentation contributes to economic and social benefits and sustainable development of families and communities [8]. Poor hygiene or improper post-handling fermentation limits shelf-life and becomes dependent on information from developed countries and technology transfer [9].

Regulators will only be convinced on causal relationships existing between fermented foods/beverages and health benefit or eventual risk, through the development of scientific dossiers, which is only feasible by industrialized producers [10].

Fermented foods may be recommended for improving the health and nutritional quality of traditional African foods and regular inclusion of fermented products as part of the daily diet would be desirable. However, lack of knowledge and understanding toward fermented food preparation may limit their usage.

Hazard Analysis and Critical Control Point (HACCP) studies in Africa of some fermented products have demonstrated that, depending on the process and the hygienic conditions observed during preparation, some fermented foods, e.g., *togwa*—fermented cereals prepared in Tanzania—may pose a safety risk [11].

Accidents may occur. For example, in 2015, at least 75 people died and some 180 fell ill, including a toddler, in the north-west of Mozambique, from apparent poisoning after consuming during a wedding party traditional fermented beer (made of sorghum, bran, corn, sugar, with *Schizosaccharomyces pombe* yeast, which belongs to the division Ascomycota, which represents the largest and most diverse group of fungi) known as "*pombe*" (Swahili word for beer). The exact cause of the contamination was later connected with bacterium *Burkholderia gladioli* and two produced toxins, bongkrekic acid and toxoflavin [12].

These safety concerns relating to pathogenic bacteria or chemical intoxicants produced by contaminating microorganisms, yeasts or moulds on a fermented food were also demonstrated by the deaths and risks of esophageal cancer reported by the consumption of fermented milk products from

Kenya including *Mursik* (a cow or goat´s milk fermented in a calabash gourd), *Kule naoto*, *Amabere amaruranu* and *Suusa* [13–15].

3. Fermented Foods and Probiotics

Fermented foods belong to a category of foods called "functional foods that are known to have a positive effect on health" [16]. Probiotics are the bacteria used to ferment traditional foods, and they are the most reported and researched. Thus, fermented foods and probiotics are closely related and co-exist despite the increased commercial interest in probiotics due to the health attributes associated with them [17]. However, the efficacy of probiotics is enhanced when taken in the form of fermented food rather than as probiotics alone [18].

Microbial food cultures have directly or indirectly come under various regulatory frameworks in the course of the last decades. Several of those regulatory frameworks put emphasis on "the history of use", "traditional food", or "general recognition of safety". Traditionally fermented foods are highly beneficial because they supply natural probiotics, now recognized as crucially important for immune health. Fermentation is an inconsistent process—almost more of an art than a science—so commercial food processors developed techniques to help standardize more consistent yields [19].

Fermentation is an anaerobic process converting sugars by bacterial enzymes to alcohol or by yeasts into lactic acid. Fermented foods are described as palatable and wholesome and are generally appreciated for several attributes: their specific unique flavors, aromas, textures, and improved cooking and processing properties. These characteristics of fermented foods are enhanced by virtue of the metabolic activities of the enzymes secreted by microorganisms [20].

Probiotics incorporate mainly fermented dairy foods. While EU permits animal production claims for feed probiotics, the USA and Canada do not. Regulators now accept several modes of action of probiotics, not just gut flora modulation [21], increasingly demanding safe strains [22]. An inventory of microorganisms used in food fermentations covering a wide range of food matrices is also required (dairy, meat, fish, vegetables, legumes, cereals, beverages, and vinegar) [23].

Various indigenous fermented foods containing probiotic bacteria have been part of local diets in Africa due to reported medicinal properties they possess [24]. Consumers must be made aware of the problems concerning raw materials and additives used in food and beverage processing as well as the possible harmful effects of employing genetically modified microorganisms in the fermentation process.

Further development of traditional fermented foods with added probiotic health features would be an important contribution towards reaching the goals of eradication of poverty and hunger, reduction in child mortality rates and improvement of maternal health in Africa [25]. Probiotic consumption may have a positive effect on psychological symptoms of depression, anxiety, and perceived stress in healthy Western human populations [26].

Fermented foods have been inching into the spotlight lately as more and more consumers learn about their inherent probiotic health benefits. The two main health effects from fermented dairy consumption are immune and metabolic positive responses, especially with the addition of probiotic organisms.

4. Public Policy: Regulations, Laws, Opinions and Guidelines

The General Food Law Regulation in the European Union (EC No. 178/2002) establishes the European Food Safety Authority (EFSA) and lays down procedures in matters of food safety, including later the principle of risk analysis.

Access to sufficient and safe food is a basic requirement for human health. In the past decades, the increased complexity of the food supply chain has contributed to the global emergence of food safety incidents. Indeed, over the past decades, a series of food safety incidents (e.g., BSE- Bovine Spongiform Encephalopathy in the 90's) have shown serious shortcomings in legal systems worldwide (e.g., law enforcement, lack of appropriate response mechanisms, control systems) [27].

It was the lack of risk communication and risk management (comprehensive/integrated approach) that brought the weaknesses of the legal food system to light and provided a new approach to food safety emerged (Figure 1) [28,29].

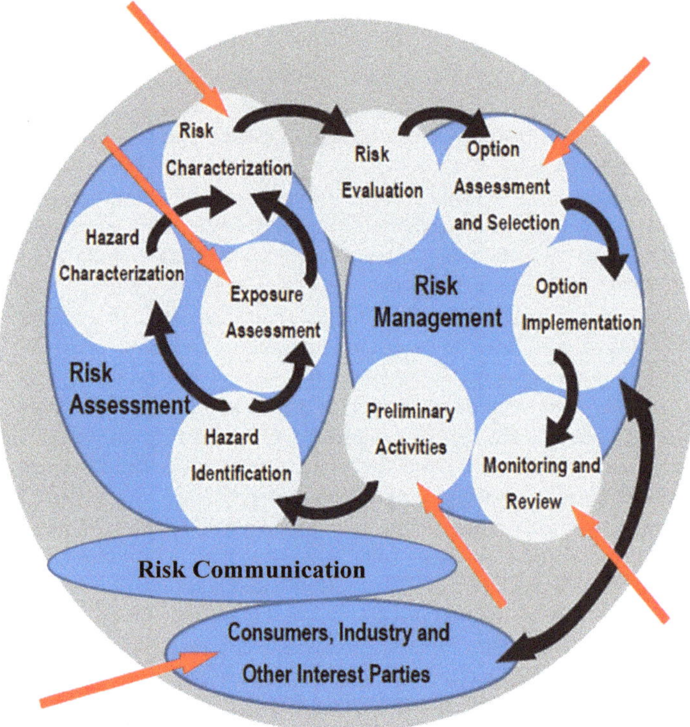

Figure 1. Integration of risk evaluation, communication and management by the EFSA.

In addition to legislation, producers of traditional fermented food and beverages are also being increasingly obliged to meet various legislative requirements. These may take various forms including taxes and "certification of origin".

The challenges of implementing measures to safeguard safety and quality in the real world are numerous. There is a notable role for ongoing training and education of all those involved as to the importance of quality and standards not only for successful trade, but also in terms of social responsibility [30].

The U.S. Food and Drug Administration (FDA) published (September 2010) the 75FR50268, *"Draft Guidance for Industry: Acidified Foods"*, which provided recommendations on manufacturing, storage, packaging, distribution processes, and appropriate quality control procedures for acid foods, acidified foods, and fermented foods. However, in December 2015 (80FR81550), they withdrew partially the draft guidance because many of the topics addressed are presently being dealt with in other documents. The FDA does not regulate fermented foods since they have not found any cases of food illnesses [31]. A comparison of key elements between EU and the USA is given below (Table 1).

Table 1. Comparison key elements of the EU–US food regulatory systems.

In Brief	EU	US
"Precautionary principle"	Fundamental part of risk management	Concept not endorsed as a basis for policy making
Societal, economic, ethical or environmental concepts	Taken into account in risk management decision in line with consumer right to information and choice	"Other factors" considered as barriers to trade
Approach to ensuring food safety	Integrated "farm-to-fork" approach	Safety mostly verified at the end of the process
Food risk evaluation	Full scientific assessment by the EFSA for regulated products such as GMO's and additives	Largely relies on companies' own private assessment

EFSA, European Food Safety Agency; GMO, Genetic Modified Organisms.

5. The Case of Fermented Soybean Extracts in Europe

Japanese cuisine, one of the healthiest foods in the world, reflected in the high life expectancy (83 years), is known to use fermented foods such as "Natto" which is sticky, smelly and slimy but still the most popular of Japan's traditional health foods.

Japan Bio Science Laboratory (JBSL) submitted in Europe a request on May 2014, under Article 4 of the Novel Food Regulation (EC No. 258/97) [32], to place on the market a fermented soybean extract (Nattokinase NSK-SD®), obtained by adding beneficial bacteria *Bacillus natto* to soybeans, as a novel food (NF). On December 2014, the competent authority of Belgium forwarded to the European Commission its initial assessment report, which came to the conclusion that the fermented soybean extract met the criteria for acceptance of a NF defined in Article 3 (1) of Regulation (EC) No. 258/79. On 6 January 2015, the Commission forwarded the initial assessment report to the other Member States (MS).

However, several of the MS submitted comments or raised objections. The concerns of a scientific nature raised by the MS on this fermented food can be summarized on their effects on hematological parameters, in particular on blood coagulation. One case report [33] was identified, in which a patient on antihypertensive agents and low-dose aspirin for secondary stroke prevention experienced an acute cerebellar hemorrhage and cerebral micro bleeds after concomitant consumption of this fermented soybean extract.

In accordance with Article 29 (1) (a) of Regulation (EC) No. 178/20024, the European Food Safety Authority (EFSA) was asked to carry out an additional assessment for the fermented soybean extract as an NF in the context of Regulation (EC) No. 258/97. The EFSA considered the elements of a scientific nature in the comments raised by MS and a Scientific Opinion was produced [34] concluding that the fermented soybean extract was safe under the intended conditions of use as specified by the applicant.

6. The Case of Gluten and Celiac Disease

Another good example is Celiac disease, a hereditary, chronic inflammatory disorder of the small intestine, which has no cure, associated with gluten intake. In the U.S. Federal Register [35], a final rule was published defining the term "gluten-free" and establishing the requirements for the voluntary use of that term in food labeling. The final rule (21 CFR 101.91) is intended to ensure that individuals with celiac disease are not misled and are provided with truthful and accurate information with respect to foods so labeled.

There is still uncertainty in interpreting the results of test methods on a quantitative basis that equates the test results to an equivalent amount of intact gluten. Therefore, alternative means are necessary to verify compliance with the provisions of the rule for fermented and hydrolyzed foods, such as cheese, yogurt, vinegar, sauerkraut, pickles, green olives, beers, and wine, or hydrolyzed plant proteins used to improve flavour or texture in processed foods such as soups, sauces, and seasonings.

7. The Homemade Foods Bill in the USA

A regulation (HB 1926) in Texas addresses genuine concerns about the risks of the food and expanded distribution. The bill addresses food safety concerns in a scale-sensitive manner, allowing for safe home food production and sales. This benefits not only producers, but also consumers, who receive improved access to healthy, locally produced foods while still providing realistic opportunity for home production, allowing home preparation of foods such as tamales, canned vegetables, fermented foods, and perishable (potentially hazardous) baked goods. Sales would be allowed anywhere in the state, including through mail order and internet sales, as long as the producer and consumer were both in Texas.

Farmers who have unsold vegetables at the end of a farmers' market may send those vegetables to their customers, and home bakers can work with grocery stores or other retail outlets to sell their goods giving a huge boost for small businesses. Home processors would be subject to regulatory provisions.

A new Homemade Food Law in California (AB 626), the largest agricultural producer and exporter in the United States, was introduced in 2017 but may ultimately be designed to meet the needs of big tech companies above the needs of home consumers and other stakeholders [36].

8. Fermented Foods and Mycotoxins

Food contamination with mycotoxins is a major problem in Africa. There is no doubt that fermentation and its products have lots of benefits. However, the fact still remains that some of the microorganisms used in the fermentation of food may become harmful under certain undesirable conditions. Microbial spores and mycotoxins such as those associated with *Aspergillus flavus*, *Aspergillus oryzae*, *Penicillium roqueforti* and other fungal toxins, such as those associated with *Fusarium*, are known to be lethal at moderate to high dosage. These toxins are produced when fermentation conditions are compromised and poor hygiene of food sources for fermentation persists during production, as happens in many African countries. *Clostridium botulinum* is an example of bacteria that causes poisoning in fermented foods and could be quite hazardous. Post fermentation contamination of products may affect the physiology of the products thereby becoming disruptive to health and deleterious to life. In order to prevent mycotoxin contamination of a fermented food, it is necessary to use a mycotoxin-free raw material and to prepare it in good sanitary and hygiene conditions [37].

9. Nutritional Guidelines and Consumer Expectations

Nutritional guidelines around the world come in many different formats, illustrated as pyramids, pie charts, text and tables, yet they are similar in terms of content [38]. Despite the wide spectrum of shapes representing food guides from around the globe, these guides use very similar methods in presenting their concepts of the ideal dietary pattern [39]. Each of these guidelines gives consumers a selection of recommended food choices (food groups) as well as a recommended daily amounts that consumers should ingest to maintain optimum health [40].

The present short review tries to provide some insights into the existing regulatory framework of Food Safety Management of some of the world's economies. Although these systems do not only differ significantly with regard to their legal culture (traditional) and legal historical background, there is a strong variation of their socio-cultural background and traditions. For now, one must exclude fermented fish from Food Guides since in parts of China an increased risk of squamous cell carcinoma of the esophagus was recorded in habitual consumers of fermented fish sauce [41] or high levels of histamine, as demonstrated in Egypt [42].

Contrary to European and USA legal systems, which are relatively young, China's legal system knows a long history that can be traced back to 563 BC based on morality. On April, 2015, the Standing Committee of China's National People's Congress revised the 2009 Food Safety Law of the People's Republic of China (Food Safety Law). The revised law came into effect on October 2015.

European Law itself is not a national legal system as Chinese Law is. Neither does it relate to a common federal law system such as in the USA. In Europe, each country has its own national legal system. In practice, discrepancies in food safety incident response can be identified in the way these legal systems deal with food safety incidents, and, as a result, in the development, design and characteristics of the regulatory framework at stake.

In general, the approach of the legal systems is broadly consistent with the assessed International (FAO/WHO) framework channeled through the Codex Alimentarius Commission (CAC) in their work to develop international food standards and guidelines [43].

Europe's food safety policy is characterized by its legal embedded preventive approach [44]. Moreover, another interesting aspect that should be taken into account in the matter of legal culture in relation to Europe is the concept of "cultural pluralism" and the characterizations thereof in the field of Food Safety Law. The "International Commission for Research into European Food History" was founded in Germany in 1989, and it deals with the history of food and nutrition in Europe since the late eighteenth century [45].

In Russia, there are "Requirements for Ferments and Enzyme Preparations" (article 12) and "Requirements as to Facilities for Ferment and Probiotic Microorganism Production" (articles 13 and 26) within a very comprehensive Federal Law [46], where the need of Fermented Food for Mandatory Certification of ready to use foods or Conformity Declaration of raw materials is clear.

10. Some National Food Guides

The Chinese dietary guidelines were first published in 1989 and revised in 2007. China uses the 'Food Guide Pagoda' (Figure 2), which is divided into five levels of recommendations.

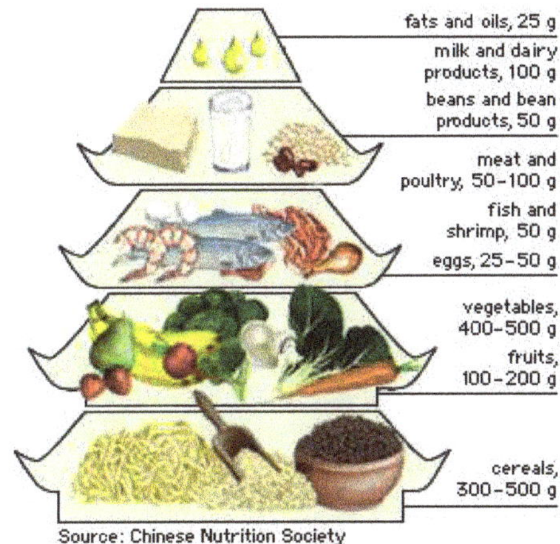

Figure 2. Food guide pagoda.

In Switzerland, in order to reduce trade barriers with the European Union, the Swiss food law has been adapted to the European law (Figure 3). Some Swiss special provisions for the reason of health protection persist. For example, a positive list of health related allegations for food and the obligation for self-supervision, and reporting and licensing requirements [47].

(A)   **(B)**

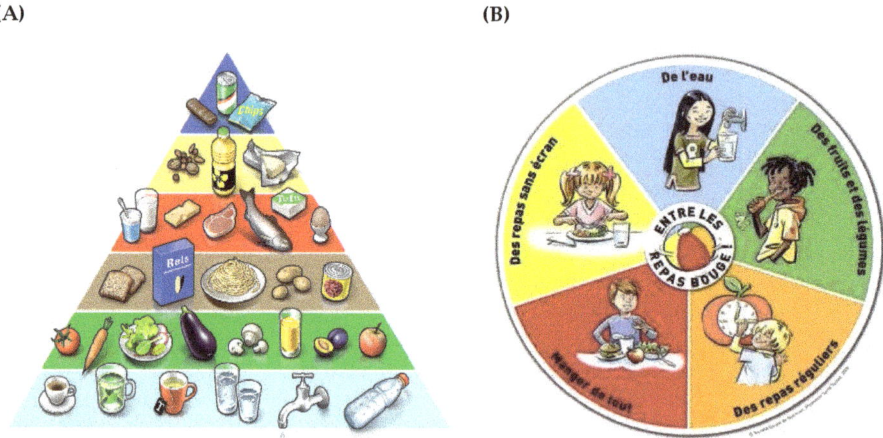

Figure 3. The Swiss food pyramid and children's nutrition disk (**A**), the pyramid; (**B**), the plate for children.

In the USA (Figure 4) and Canada (Figure 5), food guidelines mention yogurt and kefir, but there is no emphasis on them being fermented foods, nor is there inclusion of fermented foods as a healthy category [48].

Figure 4. The USA MyPyramid food guide. USDA, Unites States Department of Agriculture.

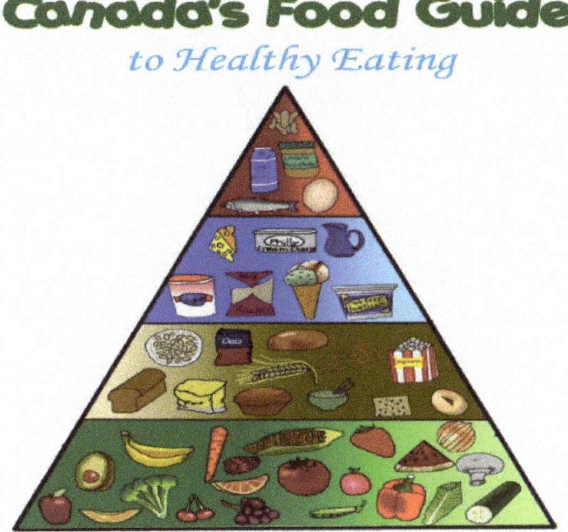

Figure 5. The Canadian food guide.

The United Kingdom (Figure 6) published its first set of dietary guidelines in 1994, and they have been regularly updated since then. Their Food Guide is still presented as a plate, and there is no category of fermented food. The "Eatwell Guide" (2017) has been recently approved [49].

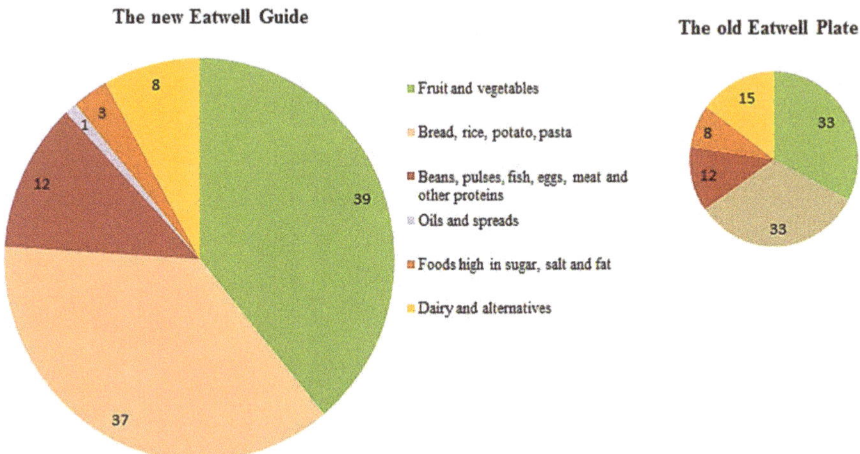

Figure 6. The UK food guide. Comparison with the previous food guide.

The Swedish and Norwegian models (Figure 7) for healthy eating, also in the form of a plate, have no section allotted to dairy products or any fermented foods [50,51].

Figure 7. The Scandinavian food guide.

The Australia New Zealand Food Authority produced Standards Code in 2014 (Figure 8), the standard related to Fermented Milk Products, including yoghurt; however, this was repealed in March 2016 (Table 2). Phytosterol, phytostanols and their esters may only be added to yoghurt in defined clear conditions of package size (200 g), percentage fat (1.5%) and total sterol added (the total plant sterol equivalents content added should be no less than 0.8 g and no more than 1.0 g per package) [52].

Figure 8. The Australian food guide.

Table 2. In fermented milk and the fermented milk portion of a food containing fermented milk, each component or parameter must comply with the value specified.

Component or Parameter	Value
Crude protein	minimum 30 g/kg
pH	maximum 4.5
Microorganisms used in the fermentation	minimum 10^6 cfu/g

In Japan, like in most countries, it is enhanced that every food group should be taken daily in moderation in order to achieve a well-balanced diet, but it does not specifically highlight fermented foods as a category (Figure 9) [53].

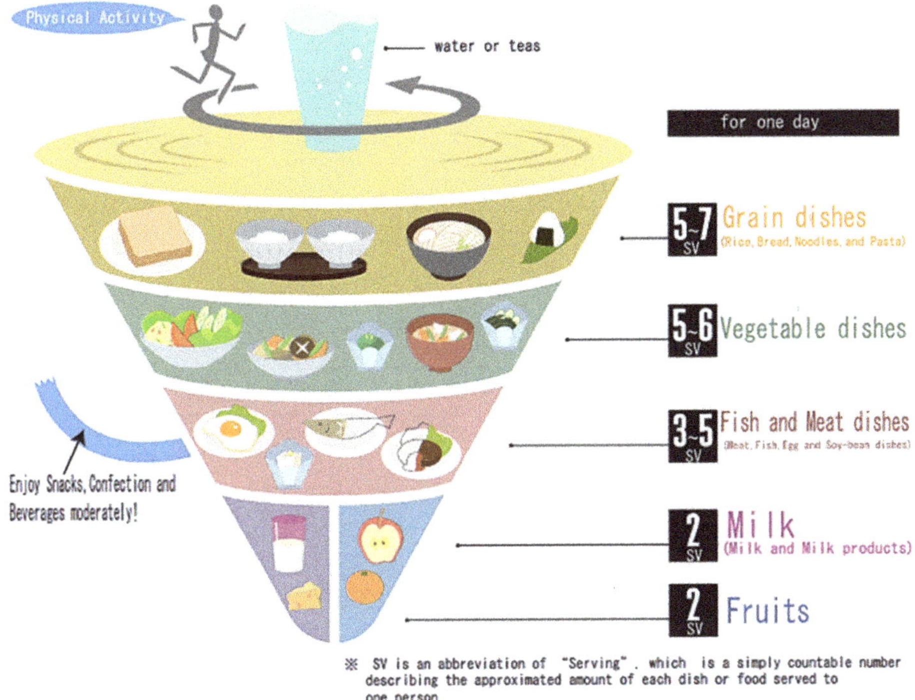

Figure 9. The Japanese food guide. SV= servings.

The one exception in Asia is India, whose pyramid of four levels explicitly encourages the consumption of fermented foods (Figure 10). The National Institute of Nutrition's 2010 "Dietary Guidelines for Indians" suggests specifically to pregnant women that they should eat more whole grains, sprouted grams and fermented foods [54].

With few exceptions, fermented foods are generally absent as a recommended category of food for daily intake in Food Guides, reflecting a failure to appreciate the benefits resulting from the process of fermentation, which have been supported by numerous studies [55].

Findings in The Netherlands, Sweden and Denmark emphasize the need to differentiate the types of dairy products, fermented and non-fermented, with regards to their health benefits, instead of promoting all dairy products, as is the case with many food guides [56–58].

The primary expectation of the general consumer today is that governments make sure proper measures are in place to ensure food sold is safe to eat. Hence, food is something we all consume; therefore, the safety of food is an issue valued and to which attention is widely drawn. To this end, several factors influence how an event is approached including the number of people ill, the severity of the illness, the distribution and volumes of food, whether the contaminant is known or unknown, and the international and trade implications.

Moreover, how people react to risks is mediated by many factors, including how risk information is perceived, how we react to social and cultural influences and how choices are structured [59], and the latter influences the legal construction designed to manage risks. What might be handled as a routine incident in one country may be considered a crisis in another [60].

Figure 10. The Indian food guide.

There are many alternatives for reinforcing healthy eating, such as the Mediterranean, Latin and Asian Pyramids. They use specific cultural eating patterns to offer evidence-based advice for healthy eating. A traditional Mediterranean diet (Figure 11), considered as the best balanced diet to follow, is high in fruits, vegetables, nuts, unrefined grains and olive oil, with a moderate intake of fish, alcohol, and a low intake of meat and dairy products, which is inversely associated with total and cardiovascular mortality [61].

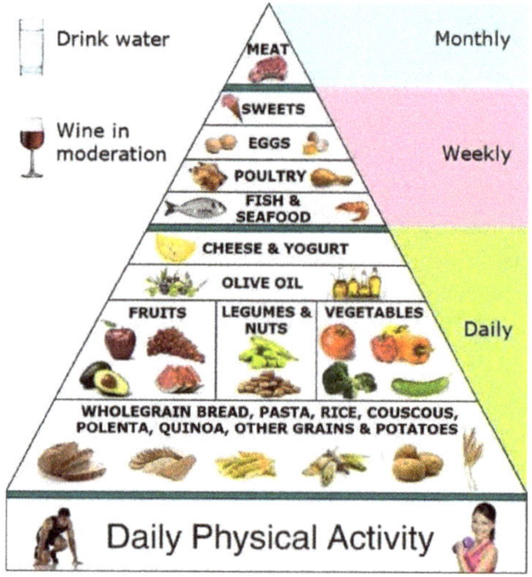

Figure 11. The Mediterranean diet pyramid.

The Portuguese Food Guide, issued in 1977, revised in 2003, is a food wheel divided into segments representing seven food groups (Figure 12). Water is in the center of the Food wheel in order to highlight the importance of hydration balance [62].

Figure 12. The Portuguese food wheel guide.

11. A Comparison of Inclusion of Fermented Products in Different Guides

Evaluating dietary guidelines from various countries can help identify their strengths and limitations, yet such assessments are lacking due to the complexity involved. Trying to compare qualitatively current population-level dietary recommendations and pictorial food guides issued by government or nutrition agencies across nutrition transition stages (early, ongoing, and transitioned) is almost an impossible task.

The recommendations for consumption of foods or dietary practices within main categories of food groups, nutrients, or beverages, including lifestyle, nutritional, cooking, or eating habits is an intricate process. Some 55 major nutrients are reasonably well characterized and their required levels of intake calculated; however, the subject becomes quite complex if taking into account the amount of phytonutrients involved in plants (over 100,000)—one of the reasons nine servings of fruits and vegetables a day are recommended. Milk or fermented milk products may block the absorption of phytonutrients, some of which are considered essential.

Below (Table 3) we attempt to provide a short comparison summary among few countries.

Table 3. Fermented products within food-based dietary guidelines in some countries.

Country	Yoghurt (Included in Dairy Products)	Alcoholic Fermented Beverages
China	-	-
Switzerland	+	in moderation
USA	+	in moderation
Canada	+	-
UK	+	-
Australia	+	in moderation
Japan	+	-
Sweden	+	-
Portugal	+	in moderation

+, existent; - non existent.

12. Conclusions and Recommendations

The present global regulatory framework is confusing and limiting. Science is developing and many exciting new technologies will continue to transform the world and improve human welfare. The market is imaginative and not discouraged by the limitations of the science.

Foods prepared by fermentation, aside from those well known in the West, will increase in amount and use spreading to other parts of the world, including the developed Western countries, as they contribute to the diversity of gut microbiota and indirect impact on mental health. The knowledge about the microbial ecology of food and beverage ecosystems is essential in order to understand the production process.

Over the past decades, the increased complexity of the food supply chain has contributed to global emergence of food safety incidents. As a result, reform of Food Safety Management regulation throughout the food supply chain has been gaining momentum in legal systems worldwide. To this end, discrepancies in food safety incident response can be identified amongst legal systems. The reasons for these discrepancies reside in different models for prediction of permissible levels, non-complete information about risks and in differences in sampling procedures. The very notion of risks, as well as the understanding and evaluation of particular risks, reflect and shape the values, preferences and prejudices of a society and therefore country differences.

The FDA in the USA and the EFSA in Europe acknowledge that many proposed rules, if finalized, may have a significant economic impact on a substantial number of small entities and may impair fair and open trade among different continents and countries. Many challenges still remain regarding the establishment of dietary guidelines integrating education, agriculture, health, environment and industry.

Author Contributions: Victoria Bell drafted and helped research the manuscript. Jorge Ferrão researched data from Africa and traditional fermented foods. Tito Fernandes conceived the idea, wrote and revised the manuscript.

Conflicts of Interest: The authors declare no conflict of interest.

References

1. Lloyd, G.E.R.; Chadwick, J.; Mann, W.N. *Regimen for Health. Hippocratic Writings*; Penguin Books Ltd.: London, UK, 1983; pp. 154–196.
2. Food and Agriculture Organization (FAO). *Traditional Fermented Food and Beverages for Improved Livelihoods*; Marshall, E., Mejia, D., Eds.; FAO: Rome, Italy, 2012.
3. Fijan, S. Microorganisms with claimed probiotic properties: An overview of recent literature. *J. Environ. Res. Public Health* **2014**, *11*, 4745–4767. [CrossRef] [PubMed]
4. Motarjemi, Y. The impact of small-scale fermentation technology on food safety in developing countries. *Int. J. Food Microbiol.* **2000**, *75*, 213–229. [CrossRef]
5. Food and Agriculture Organization (FAO). *Human Nutrition in the Developing World*; FAO: Rome, Italy, 1997.
6. Wood, B.J.B. *Microbiology of Fermented Foods*, 2nd ed.; Thomson Science: London, United Kingdom, 1998; Volume 1.
7. Zhang, J.; Wang, X.; Huo, D.; Li, W.; Hu, Q.; Xu, C.; Liu, S.; Li, C. Metagenomic approach reveals microbial diversity and predictive microbial metabolic pathways in Yucha, a traditional Li fermented food. *Sci. Rep.* **2016**, *6*, 32524. [CrossRef] [PubMed]
8. Adesulu, A.T.; Awojobi, K.O. Enhancing sustainable development through indigenous fermented food products in Nigeria. *Afr. J. Microbiol. Res.* **2014**, *8*, 1338–1343.
9. Nduko, J.M.; Matofari, J.W.; Nandi, Z.O.; Sichangi, M.B. Spontaneously fermented Kenyan milk products: A review of the current state and future perspectives. *Afr. J. Food Sci.* **2017**, *11*, 1–11.
10. Mathara, J.M.; Schillinger, U.; Kutima, P.M.; Mbugua, S.A.; Holzapfel, W.H. Isolation, identification and characterization of the dominant microorganisms of *kulenaoto*: The Maasai traditional fermented milk in Kenya. *Int. J. Food Microbiol.* **2004**, *94*, 269–278. [CrossRef] [PubMed]
11. Mugula, J.K.; Nnko, S.A.M.; Sørhaug, T. Changes in quality attributes during storage of *togwa*, a lactic acid fermented gruel. *J. Food Saf.* **2001**, *21*, 181–194. [CrossRef]

12. Mozambique (2016). Mass Poisoning Caused by Bacterial Contamination. Available online: https://www.revolvy.com/topic/Mozambique%20funeral%20beer%20poisoning. (accessed on 3 August 2017).

13. Patel, K.; Wakhisi, J.; Mining, S.; Mwangi, A.; Patel, R. Esophageal cancer, the topmost cancer at MTRH in the rift valley, Kenya, and its potential risk factors. *ISRN Oncol.* **2013**, *2013*, 503249. [CrossRef] [PubMed]

14. Nieminen, M.T.; Novak-Frazer, L.; Collins, R.; Dawsey, S.P.; Dawsey, S.M.; Abnet, C.C.; White, R.E.; Freedman, N.D.; Mwachiro, M.; Bowyer, P.; et al. Alcohol and acetaldehyde in African fermented milk Mursik—A possible etiologic factor for high incidence of esophageal cancer in Western Kenya. *Cancer Epidemiol. Biomarkers Prev.* **2013**, *22*, 69–75. [CrossRef] [PubMed]

15. Wakhisi, J.; Patel, K.; Buziba, N.; Rotich, J. Esophageal cancer in north rift valley of western Kenya. *Afr. Health Sci.* **2005**, *5*, 157–163. [PubMed]

16. Anukam, K.C.; Reid, G. African traditional fermented foods and probiotics. *J. Med. Food* **2009**, *12*, 1177–1184. [CrossRef] [PubMed]

17. Mahasneh, A.M.; Abbas, M.M. Probiotics and traditional fermented foods: The eternal connection (mini-review). *Jordan J. Biol. Sci.* **2010**, *3*, 133–140.

18. Ranadheera, R.D.; Bains, S.K.; Adams, M.C. Importance of food in probiotic efficacy. *Food Res. Int.* **2010**, *43*, 1–7. [CrossRef]

19. Shiby, V.K.; Mishra, H.N. Fermented milks and milk products as functional foods—A review. *Crit. Rev. Food Sci. Nutr.* **2013**, *53*, 482–496. [CrossRef] [PubMed]

20. Rolle, R.; Satin, M. Basic requirements for the transfer of fermentation technologies to developing countries. *Int. J. Food Microbiol.* **2002**, *75*, 181–187. [CrossRef]

21. McCartney, E. The Good, the Bad, and the Ugly—The Future Global Regulation of Probiotics for Humans and Animals. In Proceedings of the International Scientific Conference on Probiotics and Prebiotics, Budapest, Hungary, 20–22 June 2017.

22. European Commission (EC). Report of the Scientific Committee on Animal Nutrition on Product Toyocerin® for Use as Feed Additives. Available online: https://ec.europa.eu/food/sites/food/files/safety/docs/animal-feed_additives_rules_scan-old_report_out72.pdf (accessed on 3 August 2017).

23. Bourdichon, F.; Casaregola, S.; Farrokh, C.; Frisvald, J.C.; Gerds, M.L.; Hammes, W.P.; Harnett, J.; Huys, G.; Laulund, S.; Ouwehand, A.; et al. Review. Food fermentations: Microorganisms with technological beneficial use. *Int. J. Food Microbiol.* **2012**, *154*, 87–97. [CrossRef] [PubMed]

24. Food and Agriculture Organization (FAO). Current status and options for biotechnologies in food processing and in food safety in developing countries. In Proceedings of the FAO International Technical Conference, Guadalajara, Mexico, 1–4 March 2010.

25. Franz, C.M.; Huch, M.; Mathara, J.M.; Abriouel, H.; Benomar, N.; Reid, G.; Galvez, A.; Holzapfel, W.H. African fermented foods and probiotics. *Int. J. Food Microbiol.* **2014**, *190*, 84–96. [CrossRef] [PubMed]

26. Villena, J.; Kitazawa, H. Probiotic microorganisms: A closer look. *Microorganisms* **2017**. [CrossRef] [PubMed]

27. Tamang, J.P.; Shin, D.-H.; Jung, S.-J.; Chae, S.-W. Functional properties of microorganisms in fermented foods. *Front. Microbiol.* **2016**, *7*, 578. [CrossRef] [PubMed]

28. Food and Agriculture Organization/World Health Organization (FAO/WHO). *Probiotics in Food. Health and Nutrition Properties and Guidelines for Evaluation*; Food and Agriculture Organization/World Health Organization (FAO/WHO): Rome, Italy, 2006.

29. EFSA-Member State Risk Communication Handbook Updated. Available online: https://www.efsa.europa.eu/en/press/news/170524-0 (accessed on 3 August 2017).

30. The Organisation for Economic Co-Operation and Development (OECD). *SMEs, Social Entrepreneurship and Innovation*; OECD: Warsaw, Poland, 2010; Chapter 5; pp. 185–215.

31. Lofstedt, R.E.; Vogel, D. The changing character of regulation: A comparison of Europe and the United States. *Risk Anal.* **2001**, *21*, 399–416. [CrossRef] [PubMed]

32. Regulation, H.A.T. Regulation (EC) No 258/97 of the European Parliament and of the Council of 27 January 1997 concerning novel foods and novel food ingredients. *Off. J. Eur. Communities.* **1997**, *40*, 1–7.

33. Chang, Y.Y.; Liu, J.S.; Lai, S.L.; Wu, H.S.; Lan, M.Y. Cerebellar hemorrhage provoked by combined use of nattokinase and aspirin in a patient with cerebral micro bleeds. *Int. Med.* **2008**, *47*, 467–469. [CrossRef]

34. European Food Safety Authority (EFSA). *EFSA Panel on Dietetic Products, Nutrition and Allergies (NDA)*; EFSA: Parma, Italy, 2016.

35. National Archives and Records Administration. Federal Register. Available online: https://www.gpo.gov/fdsys/pkg/FR-2015-11-18/pdf/FR-2015-11-18.pdf (accessed on 3 July 2017).

36. Food Law in California; February 2017. Assembly Bill AB 626. California Retail Food Code: Microenterprise home kitchen operations. Available online: http://leginfo.legislature.ca.gov/faces/billNavClient.xhtml?bill_id=201720180AB626 (accessed on 3 July 2017).

37. Inetianbor, J.; Ezeonu, C.S.; Ezeonu, N.C.; Otitoju, O.F. Mycotoxins Associated with Fermented Foods: A Critical Review. Available online: https://www.researchgate.net/publication/288182136 (accessed on 15 June 2017).

38. Canada's Food Guides from 1942 to 1992. Available online: www.hc-sc.gc.ca/fn-an/food-guide-aliment/context/fg_history-histoire_ga-eng.php#fnb1 (accessed on 12 July 2017).

39. Painter, J.; Rah, J.-H.; Lee, Y.-K. Comparison of international food guide pictorial representations. *J. Am. Diet. Assoc.* **2002**, *102*, 483–489. [CrossRef]

40. Chilton, S.N.; Burton, J.P.; Reid, G. Inclusion of fermented foods in food guides around the world. *Nutrients* **2015**, *7*, 390–404. [CrossRef] [PubMed]

41. Ke, L.; Yu, P.; Zhang, Z.X. Novel epidemiologic evidence for the association between fermented fish sauce and esophageal cancer in South China. *Int. J. Cancer* **2002**, *99*, 424–426. [CrossRef] [PubMed]

42. Rabie, M.A.; Elsaidy, S.; el-Badawy, A.A.; Siliha, H.; Malcata, F.X. Biogenic amine contents in selected Egyptian fermented foods as determined by ion-exchange chromatography. *J. Food Prot.* **2011**, *74*, 681–685. [CrossRef] [PubMed]

43. Food and Agriculture Organization/World Health Organization (FAO/WHO). Evaluation of Certain Food Additives and Contaminants. Available online: http://www.who.int/ipcs/publications/jecfa/reports/trs940.pdf (accessed on 3 August 2017).

44. Commission Regulation (EC) No. 2073/2005 of 15 November 2005 on *Microbiological Criteria for Foodstuffs*. Available online: http://eur-lex.europa.eu/LexUriServ/LexUriServ.do?uri=OJ:L:2005:338:0001:0026:EN:PDF. (accessed on 3 August 2017).

45. ICREFH. The International Commission for Research into European Food History. 2014. Available online: http://www.vub.ac.be/SGES/ICREFH.html (accessed on 3 August 2017).

46. Russian Federal Law. 2003. "About Technical Regulating". SanPiN 2.3.2. 1290-03. "Hygienic Requirements for Manufacturing and Realization of Biological Active Additives to Food". Available online: http://www.all-certification.ru/regulation.html (accessed on 3 July 2017).

47. Di Giampaolo, A.; Capuano, V. The harmonization of Swiss and EU Food Law. A Leatherhead Food Research White Paper. Swiss Food Law Effective May 2017. Available online: https://www.leatherheadfood.com/files/2017/01/White-paper (accessed on 3 August 2017).

48. FAO. Food-Based Dietary Guidelines—Switzerland. Available online: http://www.fao.org/nutrition/education/food-based-dietary-guidelines/regions/countries/switzerland/en/ (accessed on 3 July 2017).

49. Public Health England in Association with the Welsh Government, Food Standards Scotland and the Food Standards Agency in Northern Ireland (2017). The Eatwell Guide. Available online: https://www.gov.uk/government/publications/the-eatwell-guide (accessed on 17 July 2017).

50. The National Food Agency. Uppsala, Sweden. 2013. Available online: http://www.slv.se/upload/dokument/mat/mat_skola/Good_school_meals.pdf (accessed on 3 August 2017).

51. Nordic Nutrition Recommendations. *Integrating Nutrition and Physical Activity*; Nordic Council of Ministers: Copenhagen, Danmark, 2012.

52. FSANZ. Food Standards Australia New Zealand. 2015. Available online: www.foodstandards.gov.au (accessed on 3 July 2017).

53. Yoshiike, N.; Hayashi, F.; Takemi, Y.; Mizoguchi, K.; Seino, F. A new food guide in Japan: The Japanese food guide Spinning Top. *Nutr. Rev.* **2007**, *65*, 149–154. [CrossRef] [PubMed]

54. Dietary Guidelines for Indians: A Manual. Available online: http://ninindia.org/DietaryguidelinesforIndians-Finaldraft.pdf (accessed on 13 May 2017).

55. Wilkinson, J. The Food Processing Industry, Globalization and Developing Countries. *J. Agric. Dev. Econ.* **2004**, *1*, 184–201.

56. Keszei, A.P.; Schouten, L.J.; Goldbohm, A.; van den Brandt, P.A. Dairy intake and the risk of bladder cancer in The Netherlands cohort study on diet and cancer. *Am. J. Epidemiol.* **2009**, *171*, 436–446. [CrossRef] [PubMed]

57. Sonedstedt, E.; Wirfält, E.; Wallstrom, P.; Gullberg, B.; Orho-Melander, M.; Hedblad, B. Dairy products and its association with incidence of cardiovascular disease: The Malmö diet and cancer cohort. *Eur. J. Epidemiol.* **2011**, *26*, 609–618. [CrossRef] [PubMed]

58. Adegboye, A.R.; Christensen, L.B.; Holm-Pedersen, P.; Avlund, K.; Boucher, B.J.; Heitmann, B.L. Intake of dairy products in relation to periodontitis in older Danish adults. *Nutrients* **2012**, *4*, 1219–1229. [CrossRef] [PubMed]

59. Food and Agriculture Organization/World Health Organization (FAO/WHO). *FAO/WHO Framework for Developing National Food Safety Emergency Response Plans*; United Nations: Rome, Italy, 2010.

60. Sunstein, C.R. "Choosing Not to Choose". In Proceedings of the Risk, Perception, and Response Conference, Harvard Center for Risk Analysis, Boston, MA, USA, 20–21 March 2014; Harvard School for Public Health, 2014.

61. Trichopoulou, A.; Costacou, T.; Bamia, C.; Trichopoulos, D. Adherence to a Mediterranean diet and survival in a Greek population. *N. Engl. J. Med.* **2003**, *348*, 2599–2608. [CrossRef] [PubMed]

62. Rodrigues, S.S.P.; Franchini, B.; Graça, P.; Almeida, M.D.V. A new food guide for the Portuguese population: Development and technical considerations. *J. Nutr. Educ. Behav.* **2006**, *38*, 189–195. [CrossRef] [PubMed]

Article

Optimization of Baker's Yeast Production on Date Extract Using Response Surface Methodology (RSM)

Mounira Kara Ali [1],*, Nawel Outili [2], Asma Ait Kaki [3], Radia Cherfia [1], Sara Benhassine [1], Akila Benaissa [2] and Noreddine Kacem Chaouche [1]

[1]   Laboratoire de Mycologie, de Biotechnologie et de l'Activité Microbienne (LaMyBAM), Département de Biologie Appliquée, Faculté des Sciences de la Nature et de la Vie, Université Frères Mentouri Constantine 1, Constantine 25000, Algeria; cherfiarr@yahoo.fr (R.C.); sarabenhassine@gmail.com (S.B.); nkacemchaouche@yahoo.fr (N.K.C.)

[2]   Laboratoire de Génie des Procédés de l'Environnement, Faculté du Génie des Procédés, Université Constantine 3, Constantine 25000, Algeria; out.nawel@gmail.com (N.O.); akilabenaissa@yahoo.fr (A.B.)

[3]   Département de Biologie, Faculté des Sciences de la Nature et de la Vie, Université M'hamed Bougera, Boumerdess 35000, Algeria; askaki.biotechno@gmail.com

*   Correspondence: kr.mounira@yahoo.fr; Tel.: +213-55-57-73-223

Received: 16 July 2017; Accepted: 3 August 2017; Published: 7 August 2017

Abstract: This work aims to study the production of the biomass of *S. cerevisiae* on an optimized medium using date extract as the only carbon source in order to obtain a good yield of the biomass. The biomass production was carried out according to the central composite experimental design (CCD) as a response surface methodology using Minitab 16 software. Indeed, under optimal biomass production conditions, temperature (32.9 °C), pH (5.35) and the total reducing sugar extracted from dates (70.93 g/L), *S. cerevisiae* produced 40 g/L of their biomass in an Erlenmeyer after only 16 h of fermentation. The kinetic performance of the *S. cerevisiae* strain was investigated with three unstructured models i.e., Monod, Verhulst, and Tessier. The conformity of the experimental data fitted showed a good consistency with Monod and Tessier models with R^2 = 0.945 and 0.979, respectively. An excellent adequacy was noted in the case of the Verhulst model (R^2 = 0.981). The values of kinetic parameters (Ks, X_m, μ_m, p and q) calculated by the Excel software, confirmed that Monod and Verhulst were suitable models, in contrast, the Tessier model was inappropriately fitted with the experimental data due to the illogical value of Ks (−9.434). The profiles prediction of the biomass production with the Verhulst model, and that of the substrate consumption using Leudeking Piret model over time, demonstrated a good agreement between the simulation models and the experimental data.

Keywords: *Saccharomyces cerevisiae*; biomass; date extract; optimization; response surface methodology; kinetic models

1. Introduction

For thousands of years, micro-organisms have been spontaneously used in human food preparation. However, scientists did not initiate studying these living beings until the appearance of the microscope in 1680. Among microorganisms widely studied and used in diverse biotechnological applications, the yeast *S. cerevisiae* was mentioned [1–3]. This yeast species was known formerly for its particular exploitation in the production of wine, beer and bread. Recently, it has been used as a "cell factory", able to synthesize a large spectrum of bioactive molecules as recombinant proteins, antibiotics and bioethanol [4–7]. Algeria records a remarkable lack in beet and cane molasses production, indeed, it imports about 18,000 and 13,000 tons per year of each one, respectively, and also imports the yeast strain *S. cervisiae* used as a baker's leaven [8]. However, Algeria has an enormous potential of dates [9,10]. In addition, the production of *S. cerevisiae* biomass from a low quality date variety could constitute an economic carbon source, especially considering that the production of dates is a bountiful in Algeria. The main objective of the present work is to study the optimization of *S. cerevisiae* biomass production, using date extract as a sole carbon source.

The traditional technique used for optimizing a multivariable fermentation process is difficult and does not take the alternative effects between components into consideration [11,12]. Recently, many statistical experimental design methods have been employed in bioprocess optimization [13–16]. Among these, the central composite experimental design (CCD) is the most suitable for identifying the individual variables to optimize a multivariable system [17,18]. This method was used to optimize many fermentation process, such as acids, antibiotics, enzymes, biomass and ethanol production by several micro-organisms types [19–22]. Furthermore, it was used in design, analysis, and in unit operations. The advantages of this method are the reduction of the number of experiments, reagents, time, financial input and energy [23]. The present work was conducted following these steps: (a) selecting the optimum conditions of three parameters (temperature; initial pH, and sugar concentration extracted from dates) to obtain a high yield of *S. cerevisiae* cells growth using a surface response methodology; (b) exploit the date extract as the sole carbon source for the production of *S. cerevisiae* at optimized conditions; (c) predict the biomass production process by unstructured kinetic models.

2. Materials and Methods

2.1. Origin and Reactivation of the Yeast S. cerevisiae

The yeast used in this study has a commercial origin in fact, it is produced by the factory Lesaffre. The lyophilized form of *S. cerevisiae* was chosen due to several advantages, such as its availability, rapid growth, resistance to contaminants, easy cultivation, ability to consume most sugar and high yield production. The yeast was reactivated on agar plates containing YPGA medium composed of yeast extract 10 g/L, peptone 10 g/L, glucose 20 g/L, agar 20 g/L with a pH 6, incubated at 30 °C for 24 h. After development, the yeast was analyzed by macroscopic and microscopic characteristics in order to confirm its aspect.

2.2. Preparation of Dates Extract

The date variety used in this study is very widespread in the south-eastern region of Algeria. It is the dry variety *Mech-Degla*. This variety has a sub-cylindrical shape, slightly narrowed at its tip. It is lightly beige-tinted. The epicarp is wrinkled, shiny and brittle. The mesocarp is not very fleshy of a dry consistency and fibrous texture (Figure 1) [8]. The choice of this variety is justified by its abundance at the national level, low market value, ease of preservation (dry date) and richness in sugars. This variety is therefore a favorable substrate for the yeast growth and development.

Figure 1. Dates *Mech-Degla*.

To prepare the date extract, 1 kg of this date is washed, peeled and placed in 5 L of distilled water and then boiled at 100 °C for one hour in order to extract the sugars. The obtained extract is filtered through muslin to remove the large particles, and then the solution is centrifuged at 5000 rpm for 5 min. The supernatant obtained constitutes the date extract [24].

2.3. Preparation of Culture Medium Based on the Dates Extract and Inoculums

The method cited by Kocher and Uppal [25] was used with minor modifications. The obtained date extract from the above preparation was supplemented by mineral salts: magnesium sulfate 0.44 g, urea 12.70 g, and ammonium sulfate 5.30 g. Finally, the medium was distributed in an Erlenmeyer of 250 mL with a ratio of 100 mL per flask and sterilized at 120 °C for 20 min. The pre-culture was obtained by inoculating two colonies of the yeast *S. cerevisiae* in 250 mL shake flasks containing 100 mL of dates extract, mentioned above. The pre-culture was incubated at 30 °C for 3 h, and used further as inoculums for the yeast biomass production.

2.4. Statistical Design of Experiments

2.4.1. Factor Selection and Organization of Experiments

The organization of the experiments was carried out using the experimental design obtained by the central composite experimental design (CCD). Three independent variables were selected (temperature, initial pH and concentration of sugars extracted from dates). Table 1 shows the domain of study with coded levels and real values of studied variables.

Table 1. Coded levels and real values of studied variables.

Variables	Coded Levels				
	$-\alpha$	-1	0	$+1$	$+\alpha$
	Real Values				
X_1 = Temperature (°C)	27	29	33	37	39
X_2 = Initial pH	2.4	3.6	5.5	7.3	8.6
X_3 = concentration of sugars (g/L)	1	44.1	107.5	170.9	214

In the central composite design, the -1 and $+1$ correspond to the lower and the higher level, respectively. The value 0 represents the central value of the rangeand α has the value of 1.68 ($\alpha = \sqrt[4]{N_f}$, where N_f is a number of experiments).

The CCD matrix employed for three independent variables is given in Table 2. Each column represents the different variables (factors) and each line represents the different experiments (20).

Table 2. The central composite experimental design (CCD) matrix for different variables (coded levels).

Experiments	Coded Levels		
	X_1	X_2	X_3
01	-1	-1	-1
02	$+1$	-1	-1
03	-1	$+1$	-1
04	$+1$	$+1$	-1
05	-1	-1	$+1$
06	$+1$	-1	$+1$
07	-1	$+1$	$+1$
08	$+1$	$+1$	$+1$
09	-1.68	0	0
10	$+1.68$	0	0
11	0	-1.68	0
12	0	$+1.68$	0
13	0	0	-1.68
14	0	0	$+1.68$

Table 2. *Cont.*

Experiments	Coded Levels		
	X_1	X_2	X_3
15	0	0	0
16	0	0	0
17	0	0	0
18	0	0	0
19	0	0	0
20	0	0	0

The CCD matrix is composed of a complete factorial design, 2^3; two axial points on the axis of each design variable at a distance of α = 1.682 from the design center and 5 points at the domain center. The actual experimental values corresponding to the coded levels used for the creation of the experiment matrix are presented below (Table 3).

Table 3. Actual values for the three independent variables.

Experiments	Actual Values		
	Temperature (°C)	Initial pH	Sugars Concentration (g/L)
01	29	3.6	44.1
02	37	3.6	44.1
03	29	7.3	44.1
04	37	7.3	44.1
05	29	3.6	170.9
06	37	3.6	170.9
07	29	7.3	170.9
08	37	7.3	170.9
09	27	5.5	107.5
10	39	5.5	107.5
11	33	2.4	107.5
12	33	8.6	107.5
13	33	5.5	1
14	33	5.5	214
15	33	5.5	107.5
16	33	5.5	107.5
17	33	5.5	107.5
18	33	5.5	107.5
19	33	5.5	107.5
20	33	5.5	107.5

2.4.2. Effect Estimation

The real values X have been calculated according to Equation (1).

$$X = \frac{x - x_0}{\Delta x} \tag{1}$$

Where X, is the coded value for the independent variable, x, is the natural value, x_0, is the natural value at the center point and Δx, is the step change value (the half of the interval $(-1 +1)$). The mathematical model describing the relation between dependent and independent variables for this process has the quadratic form for the experimental design used:

$$Y_i = \beta_0 + \beta_1 X_1 + \beta_2 X_2 + \beta_3 X_3 + \beta_{11} X_1^2 + \beta_{22} X_2^2 + \beta_{33} X_3^2 + \beta_{12} X_1 X_2 + \beta_{13} X_1 X_3 + \beta_{23} X_2 X_3 \tag{2}$$

where Y_i, is the predicted response (in our case, the Biomass production (g/L); β_0, is offset term; β_1, β_2, β_3 are the linear effects (showing the predicted response); β_{11}, β_{22}, β_{33} are the squared effects β_{12}, β_{13}, β_{23} are the interaction terms and X_1, X_2, X_3 are the independent variables. The calculation of the effect of each variable and the establishment of a correlation between the response Y_i and the variables X, were performed using a Minitab 16 software (Minitab, Inc., State College, PA, USA).

2.4.3. Statistical Analysis

The statistical analysis was performed using analysis of variance (ANOVA), in order to validate the square model regression. It included the following parameters: coefficient of determination R^2 Student test (t); Fisher test (F); and p-value. In our study, the statistical significance test level was set at 5% (probability (p) < 0.05).

2.5. Validation of Biomass Production in Optimum Medium

In order to confirm the optimized conditions obtained by the central composite design, an experiment was carried out on 250 mL shake flasks. To do this, 100 mL of date extract (mentioned above) was seeded with 11 mL of the yeast pre-culture and the pH of the medium was adjusted to 5.35 (optimum value). Shake flasks were sterilized at 120 °C for 20 min, and incubated at 32.9 °C (optimum value) for 18 h.

2.5. Analytical Techniques

2.5.1. Determination of Total Reducing Sugars

The total reducing sugars were determined according to the method of Dubois et al. [26], with minor modifications. The sample was filtered, and 1 mL of it was transferred to a glass tube. Then 0.6 mL of 5% (*w/v*) phenol and 3.6 mL of 98% sulfuric acid were added. The mixture was shaken and incubated at room temperature for 30 min; sugar gives a cream yellow whose intensity is proportional to the amount of total sugars. The absorbance was measured at 490 nm. A calibration curve was established using glucose as the standard.

2.5.2. Determination of Biomass Concentration

The measurement of biomass was followed by estimation of cell dry weight, expressed in g/L. one mL of yeast culture was centrifuged at 5000 rpm for 5 min. The supernatant obtained was washed twice with water and dried by incubation at 105 °C until at a constant weight [27].

2.6. Modeling

Unstructured kinetic models using Monod, Verhulst, and Tessier [28] (Table 4) have been implemented to fit the experimental data. Kinetic parameters (μ_{max}, Ks and X_m), were determined using the curve fitting method of each model. The fitness evaluation of experimental data on cell growth by models was performed using Excel software (Microsoft, Redmond, WA, USA).

Profile Prediction of Biomass and Substrate Concentration

To predict the experimental profile of biomass of *S. cerevisiae* during time fermentation, the integration of the Verhulst model was used to give a sigmoidal variation of X as a function of t, which may represent both an exponential and a stationary phase (Equation (3)):

$$X = \frac{X_0 e^{\mu_m t}}{\{1 - (X_0/X_m)(1 - e^{\mu_m t})\}} \tag{3}$$

In addition, the substrate model (Leudeking Piret) as described below (Equation (4)) was also applied to predict an experimental profile for total reducing sugars consumption by *S. cerevisiae* during time.

$$-\frac{dS}{dt} = p\frac{dX}{dt} + qX \tag{4}$$

where $p = 1/Y_{X/S}$ and q is a maintenance coefficient.

Table 4. Unstructured kinetic models to determinate the kinetic parameters.

Kinetic Models	Equations	Linearized Form	Description	Symbols
Monod	$\mu = \mu_{max}\dfrac{S}{S+K_S}$	$\dfrac{1}{\mu} = \dfrac{K_S}{\mu_{max}}\dfrac{1}{S} + \dfrac{1}{\mu_{max}}$	Monod kinetic model is a substrate concentration dependent.	μ: is the specific growth rate (h^{-1}).
Verhulst	$\mu = \mu_{max}\left(1 - \dfrac{X}{X_m}\right)$	$\mu = \mu_{max} - \dfrac{\mu_{max}}{X_m}X$	Verhulst kinetic model is an unstructured model depends on biomass concentration.	μ_{max}: is the maximum specific growth rate (h^{-1}). K_S: is the half-saturation constant (g/L). S: is the concentration in limiting substrate (g/L).
Tessier	$\mu = \mu_{max}\left(1 - e^{-K_S S}\right)$	$\ln\mu = \dfrac{1}{K_s}S + \ln\mu_{max}$	Tessier is an unstructured model for a substrate concentration dependent.	X: is the biomass concentration (g/L). X_m: is the Maximum biomass concentration (g/L).

Equation (4) is rearranged as follows:

$$-dS = p\,dX + q\int X(t)\,dt \tag{5}$$

Substituting Equation (3) in Equation (5) and integrating with initial conditions ($S = S_0$; $t = 0$) give the following Equation:

$$S = S_0 - pX_0\left\{\frac{e^{\mu_m t}}{\left\{1 - \left(\frac{X_0}{X_m}\right)(1 - e^{\mu_m t})\right\}} - 1\right\} - q\frac{X_m}{\mu_m}ln\left\{1 - \frac{X_0}{X_m}(1 - e^{\mu_m t})\right\} \tag{6}$$

3. Results and Discussion

Microbial growth is influenced by the culture medium constituents and the physico-chemical factors in particular, temperature, pH, and substrate concentration. Indeed, in the present study, the temperature, the initial pH and the concentration of the carbon source (total sugars extracted from dates) were supposed to optimize the biomass production of *S. cerevisiae* using the central composite experimental design. The biomass concentration over 16 h of fermentation varied with the change in temperature, initial pH and sugar concentration (Table 5).

Table 5. The central composite design for biomass production.

	Coded Levels			Real Values			(Y_i): Biomass (g/L)	
Experiments	X_1	X_2	X_3	Temperature (°C)	Initial pH	Concentration of Sugar (g/L)	Observed Mean Values *	Predicted Values
01	−1	−1	−1	29	3.6	44.1	24.07	23.99
02	+1	−1	−1	37	3.6	44.1	15.99	17.45
03	−1	+1	−1	29	7.3	44.1	25.70	27.80
04	+1	+1	−1	37	7.3	44.1	15.79	20.98
05	−1	−1	+1	29	3.6	170.9	28.40	25.05
06	+1	−1	+1	37	3.6	170.9	29.86	29.59
07	−1	+1	+1	29	7.3	170.9	20.78	21.16
08	+1	+1	+1	37	7.3	170.9	23.51	25.42
09	−1.68	0	0	27	5.5	107.5	22.61	24.06
10	+1.68	0	0	39	5.5	107.5	26.20	22.15
11	0	−1.68	0	33	2.4	107.5	26.00	28.21
12	0	+1.68	0	33	8.6	107.5	32.72	27.90
13	0	0	−1.68	33	5.5	1	25.37	21.09
14	0	0	+1.68	33	5.5	214	24.04	25.71
15	0	0	0	33	5.5	107.5	40.00	40.07
16	0	0	0	33	5.5	107.5	40.00	40.07
17	0	0	0	33	5.5	107.5	40.00	40.07
18	0	0	0	33	5.5	107.5	40.00	40.07
19	0	0	0	33	5.5	107.5	40.00	40.07
20	0	0	0	33	5.5	107.5	40.00	40.07

* Each experiment was carried out twice and the average value is used here.

Using the results obtained in diverse experiments, the correlation gives the influence of temperature (X_1), initial pH (X_2) and total sugar concentration (X_3) on the response. This correlation is obtained by Minitab 16 software and expressed by the following second order polynomial (Equation (7)):

$$Y_i = 40.074 - 0.568X_1 - 0.090X_2 + 1.373X_3 - 5.999X_1{}^2 - 4.248X_2{}^2 - 5.893X_3{}^2 - 0.070X_1X_2 + 2.772X_1X_3 - 1.925X_2X_3 \tag{7}$$

Table 6 shows the coefficient regression corresponding with t and p-values for all the linear, quadratic and interaction effects of parameters tested. A positive sign in the t-value indicates a synergistic effect, while a negative sign represents an antagonistic effect of the parameters on the biomass concentration [29].

Table 6. Estimated regression coefficients of t and *p*-values of the model.

Terms	Coefficients	Square Error	*t*-Value	*p*
β_0	40.0744	1.3912	28.806	0.000
β_1	−0.5684	0.9230	−0.616	0.552
β_2	−0.0907	0.9230	−0.098	0.924
β_3	1.3739	0.9230	1.488	0.167
β_{11}	**−6.0000**	**0.8985**	**−6.677**	**0.000**
β_{22}	**−4.2481**	**0.8985**	**−4.728**	**0.001**
β_{33}	**−5.8939**	**0.8985**	**−6.559**	**0.000**
β_{12}	−0.0700	1.2060	−0.058	0.955
β_{13}	**2.7725**	**1.2060**	**2.299**	**0.044**
β_{23}	−1.9250	1.2060	−1.596	0.142

$R^2 = 91.1\%$, R^2 (adj) = 83.16%, S = 3.41104, PRESS = 884.951.

The examination of Table 6 shows that all coefficient regression of the quadratic terms are statistically significant $p \leq 0.05$ and negatively affect the biomass production (Figure 2). In contrast all coefficient regression of linear and interaction terms were statistically not significant $p > 0.005$, except the interaction term X_1X_3, which is significant $p = 0.044$ and has a synergistic effect on the response (Figure 2).

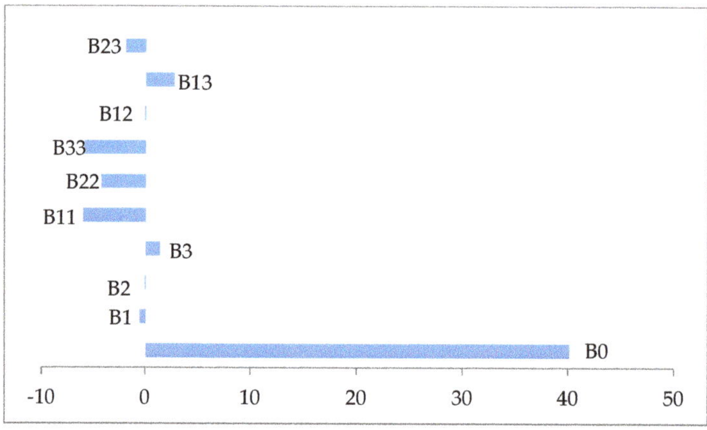

Figure 2. Variable effect signification on a biomass production.

The analysis of variance (ANOVA) of the coefficient regression for the cell growth production (Table 7) demonstrates that the model is significant due to the *F*-value of 11.43 and the low probability *p* value ($p = 0.000$). Generally, the *F*-value with a low probability *p*-value indicates a high significance of the regression model [30].

Moreover, the coefficient of determination (R^2) measures the fit between the model and experimental data. Figure 3 was also determined to evaluate the regression model. In this study, the obtained value of R^2 is 0.911 approximate to 1, which justifies an excellent consistency of the model [31]. On the other hand, the obtained R^2 implies that 91.1% of the sample variation in the cell growth is attributed to the independent variables. This value indicates also that only 8.86% of the variation is not explained by the model.

Table 7. Analysis of variance (ANOVA).

Source	DF	Seq SS	Adj SS	Adj MS	F	p
Regression	9	1196.65	1196.65	132.961	**11.43**	**0.000**
Linear	3	30.30	30.30	10.101	0.87	0.489
A	1	4.41	4.41	4.412	0.38	0.552
B	1	0.11	0.11	0.112	0.01	0.924
C	1	25.78	25.78	25.779	2.22	0.167
Square	3	1075.17	1075.17	358.390	30.80	0.000
A*A	1	379.27	518.80	518.799	44.59	0.000
B*B	1	195.28	260.07	260.071	22.35	0.001
C*C	1	500.62	500.62	500.618	43.03	0.000
Interaction	3	91.18	91.18	30.393	2.61	0.109
A*B	1	0.04	0.04	0.039	0.00	0.955
A*C	1	61.49	61.49	61.494	5.29	0.044
B*C	1	29.64	29.64	29.645	2.55	0.142
Residual Error	10	116.35	116.35	11.635		

DF: degrees of freedom; Seq SS: sequential sum of squares; Adj SS: adjusted, sum of squares; AdjMS: adjusted, mean of squares F: Fischer's variance ratio; P: probability value.

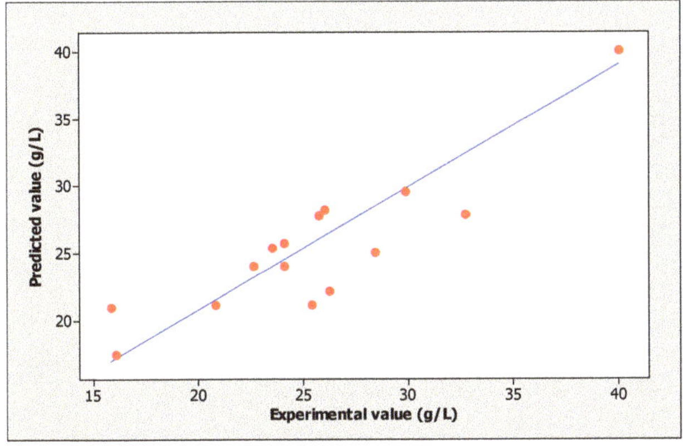

Figure 3. The fit between the model and experimental data of cell growth.

According to the literature, the study proposed by Boudjemaet al. [22] was carried out using a design of experiment to describe the batch fermentation of bioethanol and biomass production on sweet cheese whey by *Saccharomyces cerevisiae* DIV13-Z087C0VS. The results showed a good agreement with experimental data (a low probability p value ≤ 0.000 and a good correlation coefficient ($R^2 = 0.914\%$), which confirms a high significance of the regression model. In addition, the study carried out by Bennamoun et al. [32] showed that the optimization of the medium components, which enhance the polygalacturonase activity of the strain *Aureobasidium pullulans*, was achieved with the aid of the same method used in the present study (response surface methodology). The obtained results showed a significance of the method used in comparison with the experimental data; a very low p value (0.001) and a high coefficient of determination ($R^2 = 0.9421$).

The optimization of the response Y_i (Biomass production) and the prediction of the optimum levels of temperature, initial pH and sugars concentration of fermentation were obtained. This optimization resulted in surface plots (Figure 4) and an isoresponse contour plot (Figure 5).

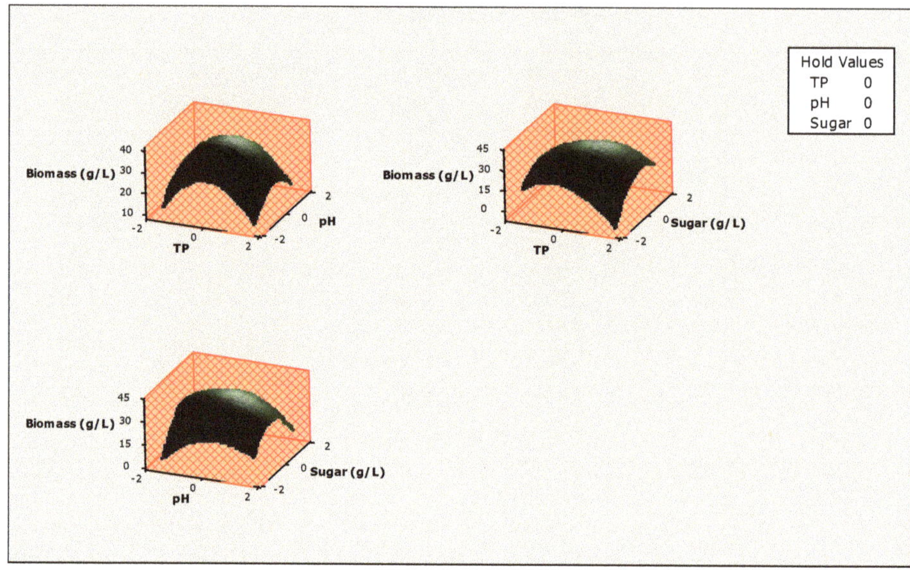

Figure 4. Surface plot for the effect of different parameters on biomass production.

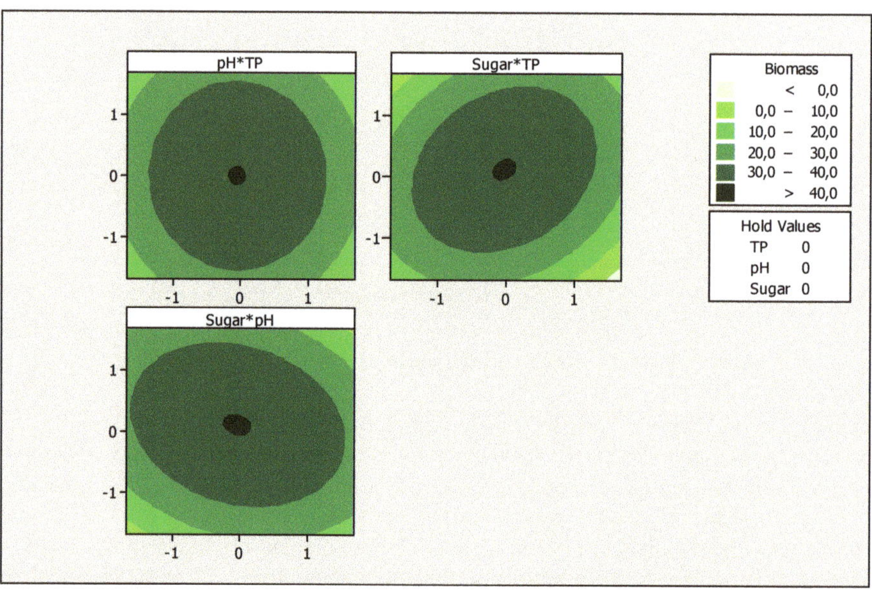

Figure 5. Isoresponse contour plot for the effect of the studied variables on biomass production.

These figures show that there is an optimum, located at the center of the field of study. In addition, the use of the minitab optimizer will give exact values of the optimum operating conditions of the process (Figure 6).

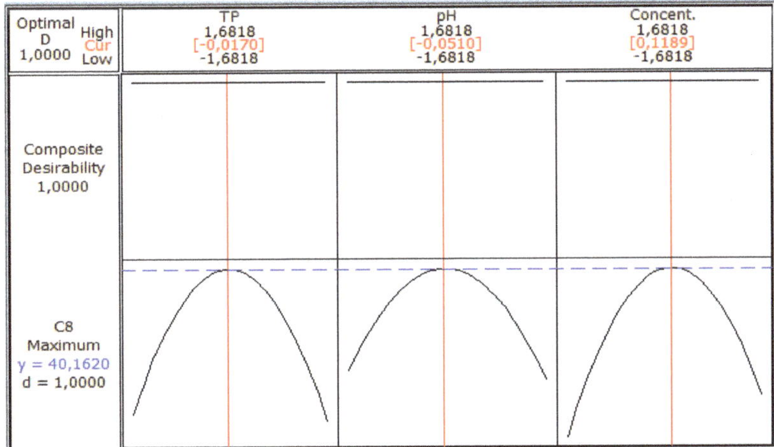

Figure 6. Coded values of optimal conditions on biomass production.

Figure 6 shows the maximum biomass production by *S. cerevisiae* (40.162 g/L) corresponding to coded values of temperature (−0.0170), pH (−0.0510) and sugar concentration (0.1189).These values are equivalent to real values of 32.9 °C, 5.35 and 70.93 g/L, respectively. Jiménez Islas et al. [27] obtained the highest cell concentration of *S. cerevisiae* ATCC 9763 (7.9 g/L) after 26 h when the strain grew at 30 °C and pH 5.5.

The validation of the baker's yeast biomass concentration and total reducing sugar consumption, over time, at optimized conditions, are presented in Figure 7. In the beginning, the biomass concentration increased with a decrease in the sugar level, reaching the maximum (40 g/L) at 16 h of fermentation, which confirms the biomass obtained by the CCD predictions (40.1620) (Figure 6). After this period, the diminution of a biomass concentration was observed, which could be explained by the sugar consumption, which ran out after18 h of fermentation.

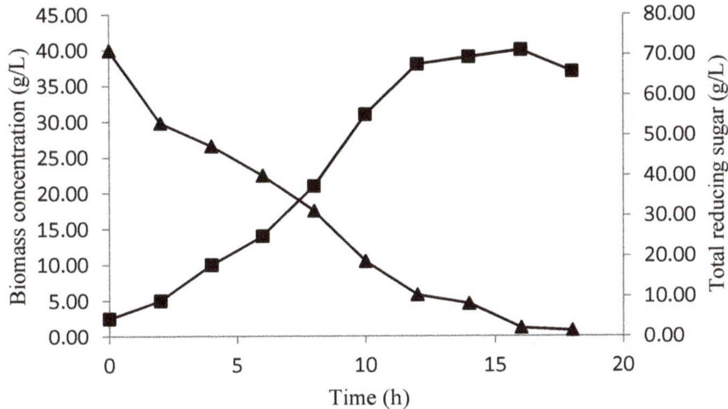

Figure 7. The biomass production (■), and total reducing sugar consumption (▲) over time at optimized conditions.

The same results were obtained by Nancib et al. [33], where the production of biomass from baker's yeast *S. cerevisiae* on a medium containing date byproducts was 40 g/L. Khan et al. [34] used six different strains of *S. cerevisiae* in fermentation medium containing date extract (with 60% sugars), in addition to 2 g/L ammonium sulfate and 50 mg/L biotin. Their results showed that the theoretical yields were about 42.8%. In addition, Al Obaidi et al. [35] studied two substrates i.e, date syrup and molasses for the propagation of baker's yeast strain *S. cerevisiae* on a pilot plant scale. The results showed that higher productivity of baker's yeast was observed when date extract was used. Other results were obtained in several studies using an alternative substrate of fermentation. In fact, the optimal biomass production (6.3 g/L) was depicted at 24 h using *Saccharomyces cerevisiae* DIV13-Z087C0VS on a medium containing sweet cheese as a sole carbon source [22]. On the other hand, the production of baker's yeast from apple pomace gives a yield of 0.48 g/g [36]. Therefore, it was concluded from these studies that the medium containing the date extract as a sole carbon source is an excellent fermentation medium for baker's yeast production.

The results of the kinetic parameters of *S. cerevisiae* growth with the different kinetic models based on the curve fitting method are presented in Table 8.

Table 8. Kinetic parameters of *S. cerevisiae* growth and substrate utilization using unstructured models.

Kinetic Models	Parameters Estimation			
	R^2	Ks (g/L)	μ_{max} (h^{-1})	X_m
Monod	0.945	0.228	0.496	-
Verhulst	0.981	-	0.376	15.04
Tessier	0.979	−9.434	0.408	

The curve fitting of cell growth using the Monod model ($1/\mu$ versus $1/S$) is presented in Figure 8. Based on the results obtained in Table 8 for this model the μ_{max} and Ks were evaluated as 0.496 h^{-1} and 0.228 g/L, respectively. These values indicate a rapid cell growth due to the high value of the specific growth rate and an elevated affinity between substrate consumption and cell growth thanks to the small half-saturation constant. In this case, R^2 was also fitted on 0.945. According to the results obtained, the Monod kinetic model is an appropriate model to make the kinetic performance of this strain.

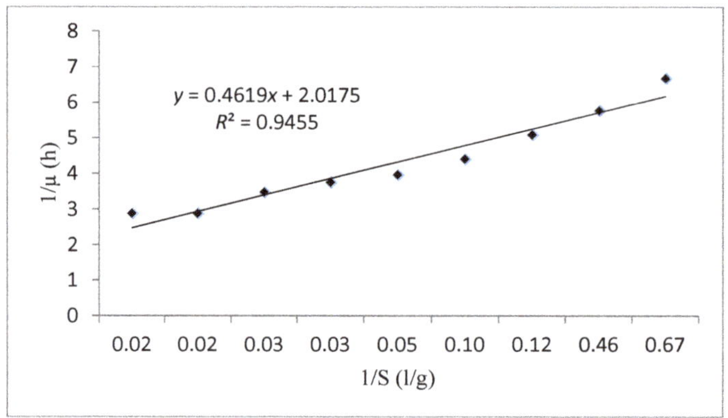

Figure 8. The Lineweaver Burk linear plot fitting the experimental data using the Monod kinetic model.

Figure 9 illustrates the linear curve fitting (μ versus X) to examine the reliability of cell kinetic performance via the Verhulst model. The analysis of the results obtained showed that the experimental data of the cell growth and substrate consumption in batch system have an excellent fitness with this model (R^2 = 0.981). The maximum specific growth rate (μ_{max}) and the maximum concentration of biomass (X_m), were 0.376 h^{-1} and 15.04 g/L respectively (Table 8). Higher values of these parameters indicated a rapid growth of the biomass which confirms the goodness of fit of the Verhulst model.

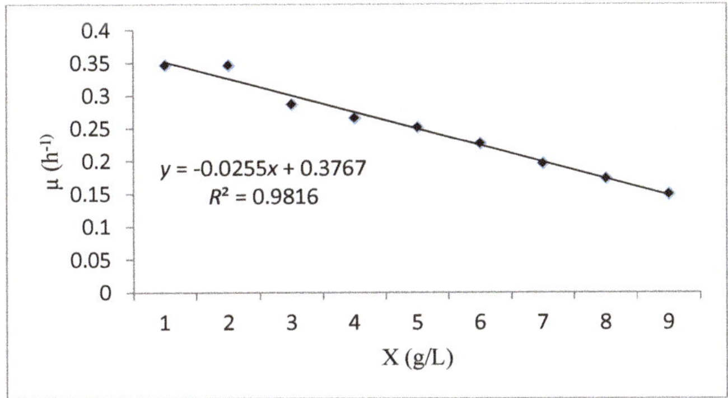

Figure 9. A plot fitting the experimental data using the Verhulst kinetic model.

The kinetic behavior fitness of *S. cerevisiae* with the Tessier kinetic model is illustrated in Figure 10. The coefficient of correlation R^2 equal to 0.979 and the estimation parameters (μ_{max} and Ks) shown in Table 8 were 0.408 and -9.434 respectively. The examination of the cell growth fitting curve with the Tessier kinetic model showed that, even though they were appropriate R^2 and μ_{max} values, the model is not suitable with the experimental data due to the illogical value of the half-saturation constant (negative Ks).

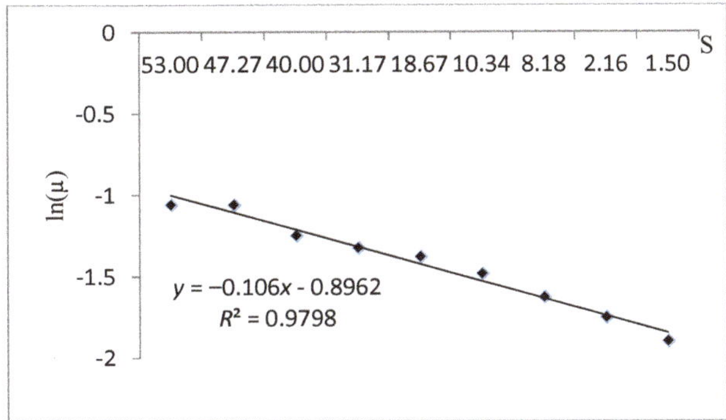

Figure 10. A plot fitting the experimental data using the Tessier kinetic model.

The comparison between the three kinetic models tested in this study showed that the Verhulst kinetic model with $R^2 = 0.981$ was the best and most appropriate model to explain *S. cerevisiae* growth and substrate utilization. Approximate results were obtained by Ardestani and Shafiei [37], who proved that the Verhulst kinetic model with R^2 equal to 0.97 was the most appropriate to describe the biomass growth rate of *S. cerevisiae*. In contrast, Ardestani and Kasebkar [38], applied an unstructured kinetic model of *Aspergillus niger* growth and substrate uptake in a submerged batch culture and have confirmed that Monod and Verhulst kinetic models were not in an acceptable range to fit a growth of *Aspergillus niger*.

A profile of biomass and total reducing sugar concentration during fermentation time is compared to the values predicted by the equations model obtained in Figure 11.

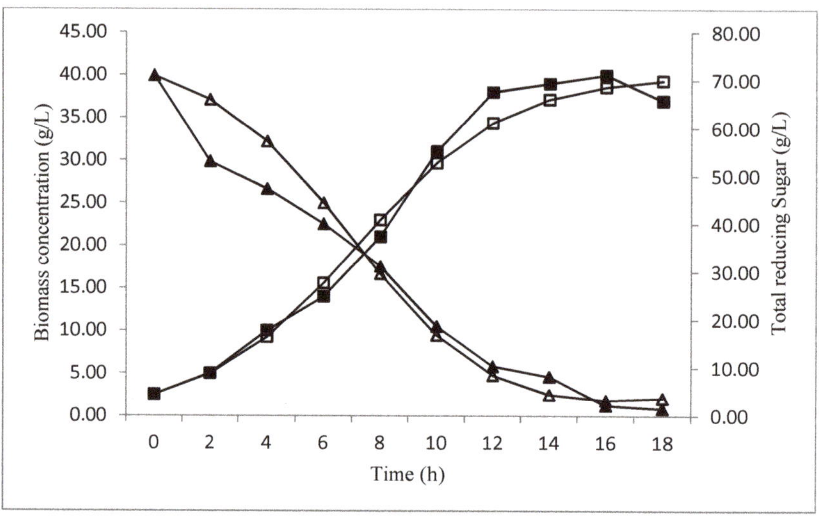

Figure 11. The comparison between predicted (□), experimental data (■) for biomass production of baker's yeast; and predicted (Δ), experimental data (▲), for total reducing sugar consumption.

At the beginning of the fermentation, values of biomass between predicted and experimental data were approximately the same. However, after 10 h and until the end of the fermentation, the difference was remarkable. In fact, the values relative to biomass were inferior compared to the values predicted by the Verhulst model. The correlation coefficient is 0.992. As for total reducing sugar concentration, the values obtained by the Leudeking Piret model were lower than those predicted in the first 7 h only. After this period, total reducing sugar values were almost identical. The correlation coefficient is 0.984. In addition, the parameter values of p and q were optimized using the experimental data for substrate based on the square minimized between observed and predicted data. Excel software illustrated the values of p = 2.1235 and q = −0.0256 h^{-1}. On the basis of these results, good correlation coefficients showed that the proposed Verhulst model and the Luedeking Piret model were adequate to explain the development of biomass production process on date extract. According to the literature, the study proposed by Kara Ali et al. [39] was carried out using the logistic empirical kinetic model and Leudeking Piret model to describe batch fermentation of *P. caribbica* on inulin. The results showed a good agreement with the experimental data (R^2 = 0.91) for cell growth and (R^2 = 0.95) for substrate consumption. In addition, the values of p and q were 14.735 and −0.077 1/h, respectively, thus, the model equations were found to represent an appropriate kinetic model for successfully describing yeast cell growth in batch fermentation. Another kinetic study proposed by Zajšek and Goršek [40] which used the unstructured models of batch kefir fermentation kinetics for ethanol production by mixed natural microflora confirmed that the growth of kefir grains could be expressed by a logistic function model, and it can be employed for the development and optimization of bio-based ethanol production processes. Furthermore, the study of Pazouki et al. [41] which illustrated the kinetic models of cell growth, substrate utilization and bio-decolorization of distillery waste water by *Aspergillus fumigatus* UB260. This study confirmed that the Logistic equation for the growth and the Leudeking Piret kinetic model for substrate utilization were able to fit the experimental data (R^2 = 0.984). The coefficient equation were also calculated (p and q) their values were 1.41 (g/g) and 0.0007 (1/h) respectively.

4. Conclusions

Microbial fermentation is complex and it is quite difficult to understand its complete details process. The central composite design (CCD) proposed in this study seems pertinent to describe the optimum biomass production of *S. cerevisiae*. A second order polynomial model was developed to evaluate the quantitative effects of temperature, pH and reducing sugar concentration in order to discover the optimum conditions for the biomass production from date extract. According to the experimental results, a maximum biomass concentration of 40 g/L was obtained at the optimum condition of temperature (32.9 °C), pH (5.35) and total reducing sugars (70.93 g/L).

In addition, among three unstructured kinetic models, both Monod and Verhulst models represent the experimental data of biomass production kinetics; nevertheless, the Verhulst model was the most suitable model to signify the baker's yeast production on date extract medium.

Acknowledgments: This work was financially supported by University Frères Mentouri Constantine 1. The authors are very grateful to the students; MALKI R. and LEKIKOT Z. (UMC-1, Algeria) for the collaboration and assistance to carry out this study.

Author Contributions: M.K. performed the experiments, wrote the paper; N.O. and A.B. helped in the statistical analysis and modeling; A.A.K and S.B. helped in interpretation and analysis of data, C.R. revised English; N.K.C. supervised the activities, contributed laboratory, reagents and materials.

References

1. Pandey, A.; Soccol, C.R.; Selvakumar, P.; Soccol, V.T.; Krieger, N.; Fontana, J.D. Recent developments in microbial inulinases, its production, properties and industrial applications. *Appl. Biochem. Biotechnol.* **1999**, *81*, 35–52. [CrossRef]

2. Richelle, A.; Ben Tahar, I.; Hassouna, M.; Bogaerts, P.H. Macroscopic modelling of bioethanol production from potato peel wastes in batch cultures supplemented with inorganic nitrogen. *Bioprocess Biosyst. Eng.* **2015**, *38*, 1819–1833. [CrossRef] [PubMed]

3. Tang, L.; Wang, W.; Zhou, W.; Cheng, K.; Yang, Y.; Liu, M.; Cheng, K.; Wang, W. Three-pathway combination for glutathione biosynthesis in *Saccharomyces cerevisiae*. *Microb. Cell Fact.* **2015**, *14*, 139. [CrossRef] [PubMed]

4. Lin, Y.; Tanaka, S. Ethanol fermentation from biomass resources: current state and prospects. *Appl. Microbiol. Biotechnol.* **2006**, *69*, 627–642. [CrossRef] [PubMed]

5. Martin, C.; Galbe, M.; Wahlbom, C.F.; Hahn-Hagerdal, B.; Jonsson, L.F. Ethanol production from enzymatic hydrolysates of sugarcane bagasse using recombinant xylose utilizing *Saccharomyces cerevisiae*. *Enzyme Microb. Technol.* **2012**, *31*, 274–282. [CrossRef]

6. Neagu, C.B.; Bahrim, G. Comparative study of different methods of hydrolysis and fermentation for bioethanol obtaining from inulin and inulin rich feedstock. *Food Ind.* **2012**, *13*, 63–68.

7. Mussato, S.I.; Machado, E.M.S.; Carneiro, L.M.; Teixeira, J.A. Sugar metabolism and ethanol production by different yeast strains from coffee industry wastes hydrolysates. *Appl. Energy* **2012**, *92*, 763–768. [CrossRef]

8. Chibane, H.; Benamara, S.; Noui, Y.; Djouab, A. Some Physichemical and Morphological Characterizations of Three Varietes of Algerian Common Dates. *Eur. J. Sci. Res.* **2007**, *18*, 134–140.

9. Kacem-Chaouche, N.; Dehimat, L.; Meraihi, Z.; Destain, J.; Kahlat, K.; Thonart, P.H. Decommissioned dates: chemical composition and fermentation substrate for the production of extracellular catalase by an *Aspergillus phoenicis* mutant. *Agric. Biol. J. N. Am.* **2013**, *4*, 41–47. [CrossRef]

10. Peyron, G. *Cultiver le Palmier Dattier*; Technical Centre for Agricultural and Rural Cooperation (CTA): Wageningen, The Netherlands, 2000; p. 110.

11. Alexeeva, Y.V.; Ivanova, E.P.; Bakunina, I.Y.; Zvaygintseva, T.N.; Mikhailov, V.V. Optimization of glycosidases production by *Pseudoalteromonasissachenkonii*. KMM 3549T. *Lett. Appl. Microbiol.* **2002**, *35*, 343–346. [CrossRef] [PubMed]

12. Patidar, P.; Agrawal, D.; Banerjee, T.; Patil, S. Chitinase production by *Beauveriafelina*. RD 101: Optimization of parameters under solid substrate fermentation conditions. *World J. Microbiol. Biotechnol.* **2005**, *21*, 93–95. [CrossRef]

13. Rajendhran, J.; Krishnakumar, V.; Gunasekaran, P. Optimization of a fermentation medium for the production of *Penicillin G* acylase from *Bacillus* sp. *Lett. Appl. Microbiol.* **2002**, *35*, 523–527. [CrossRef] [PubMed]

14. Schmidt, F.R. Optimization and scale up of industrial fermentation processes. *Appl. Microbiol. Biotechnol.* **2005**, *68*, 425–435. [CrossRef] [PubMed]

15. Singh, B.; Satyanarayana, T.A. Marked enhancement in phytase production by a thermophilic mould *Sporotrichum thermophile* using statistical designs in a cost-effective cane molasses medium. *J. Appl. Microbiol.* **2006**, *101*, 344–352. [CrossRef] [PubMed]

16. Sayyad, S.A.; Panda, B.P.; Javed, S.; Ali, M. Optimization of nutrient parameters for lovastatin production by *Monascus purpureus* MTCC 369 under submerged fermentation using response surface methodology. *Appl. Microbiol. Biotechnol.* **2007**, *73*, 1054–1058. [CrossRef] [PubMed]

17. Kennedy, M.; Krouse, D. Strategies for improving fermentation medium performance: A review. *J. Ind. Microbiol. Biotechnol.* **1999**, *23*, 456–475. [CrossRef]

18. Chakravarti, R.; Sahai, V. Optimization of compaction production in chemically defined production medium by *Penicilium citrinum* using statistical methods. *Process Biochem.* **2002**, *38*, 481–486. [CrossRef]

19. Kar, B.; Banerjee, R.; Bhattachaaryya, B.C. Optimization of physicochemical parameters for gallic acid production by evolutionary operation-factorial design techniques. *Process Biochem.* **2002**, *37*, 1395–1401. [CrossRef]

20. Naveena, B.J.; Altaf, M.; Bhadriah, K.; Reddy, G. Selection of medium components by Plackett-Burman design for production of L(+) lactic acid by *Lactobacillus amylophilus* GV6 in SSF using wheat bran. *Bioresour. Technol.* **2005**, *96*, 485–490. [CrossRef] [PubMed]

21. Shafaghat, H.; Najafpour, G.D.; Rezaei, P.S.; Sharifzadeh, M. Optimal growth of *saccharomyces cerevisiae* (ptcc 24860) on pretreated molasses for the ethanol production: The application of the response surface methodology. *Chem. Ind. Chem. Eng. Q.* **2010**, *16*, 199–206. [CrossRef]

22. Boudjema, K.; Fazouane-naimi, F.; HellaL, A. Optimization of the Bioethanol Production on Sweet Cheese Whey by *Saccharomyces cerevisiae* DIV13-Z087C0VS using Response Surface Methodology (RSM). *Rom. Biotech. Lett.* **2015**, *20*, 10814–10825.

23. Montgomery, D.C. *Design and Analysis of Experiments*, 5th ed.; John Wiley & Sons, Inc.: New York, NY, USA, 2009.

24. Bouhadi, D.; Raho, B.; Hariri, A.; Benattouche, Z.; Sahnouni, F.; Ould Yerou, K.; Bouallam, S.A. Utilization of Date Juice for the Production of Lactic Acid by *Streptococcus Thermophilus*. *J. Appl. Biotechnol. Bioeng.* **2017**, *3*, 00071.

25. Kocher, G.S.; Uppal, S. Fermentation variables for the fermentation of glucose and xylose using *Saccharomyces cerevisiae* Y-2034 and *Pachysolan tannophilus* Y-2460. *Indian J. Biotechnol.* **2013**, *12*, 531–536.

26. Dubois, M.; Gilles, K.A.; Hamilton, J.K.; Rebers, P.A.; Smith, F. Colorimetric Method for Determination of Sugars and Related Substances. *Anal. Chem.* **1956**, *28*, 350–356. [CrossRef]

27. Jiménez-Islas, D.; Páez-Lerma, J.; Soto-Cruz, N.O.; Gracida, J. Modelling of Ethanol Production from Red Beet Juice by *Saccharomyces cerevisiae* under Thermal and Acid Stress Conditions. *Food Technol. Biotechnol.* **2014**, *52*, 93–100.

28. Juska, A. Minimal models of growth and decline of microbial populations. *J. Theor. Biol.* **2011**, *269*, 195–200. [CrossRef] [PubMed]

29. Le Man, H.; Behera, S.K.; Park, H.S. Optimization of operational parameters for ethanol production from Korean food waste leachate. *Int. J. Environ. Sci. Tech.* **2010**, *7*, 157–164. [CrossRef]

30. Rene, E.R.; Jo, M.S.; Kim, S.H.; Park, H.S. Statistical analysis of main and interaction effects during the removal of BTEX mixtures in batch conditions using wastewater treatment plant sludge microbes. *Int. J. Environ. Sci. Technol.* **2007**, *4*, 177–182. [CrossRef]

31. Annuar, M.S.M.; Tan, I.K.P.; Ibrahim, S.; Ramachandran, K.B. A Kinetic Model for Growth and Biosynthesis of Medium-Chain-Length Poly-(3-Hydroxyalkanoates) in *Pseudomonas putida*. *Braz. J. Chem. Eng.* **2008**, *25*, 217–228. [CrossRef]

32. Bennamoun, L.; Hiligsmann, S.; Dakhmouche, S.; Ait-Kaki, A.; Labbani, F.Z.K.; Nouadri, T.; Meraihi, Z.; Turchetti, B.; Buzziniand, P.; Thonart, P. Production and Properties of a Thermostable, pH—Stable Exo-Polygalacturonase Using *Aureobasidium pullulans* Isolated from Saharan Soil of Algeria Grown on Tomato Pomace. *Foods* **2016**, *5*, 72. [CrossRef] [PubMed]

33. Nancib, N.; Nancib, A.; Boudrant, J. Use of waste date products in the formation of baker's yeast biomass by *Saccharomyces cerevisiae*. *Bioresour. Technol.* **1997**, *60*, 67–71. [CrossRef]

34. Khan, J.A.; Abulnaja, K.O.; Kumosani, T.A.; Abou-Zaid, A.A. Utilization of Saudi date sugars in production of baker's yeast. *Bioresour. Technol.* **1995**, *53*, 63–66. [CrossRef]

35. Al Obaidi, Z.S.; Mohamed, N.A.; Hasson, N.A.; Jassem, M.A. Semi-Industrial production of Baker's yeast using date extract and molasses. *J. Agric. Water Resour. Res.* **1986**, *5*, 162–174.
36. Bhushan, S.; Joshi, V.K. Baker's yeast production under fed-batch culture from apple pomace. *J. Sci. Ind. Res.* **2006**, *65*, 72–76.
37. Ardestani, F.; Shafiei, S. Non-Structured Kinetic Model for the Cell Growth of *Saccharomyces cerevisiae* in a Batch Culture. *Iranica J. Energy Environ.* **2014**, *5*, 8–12. [CrossRef]
38. Ardestani, F.; Kasebkar, R. Non-Structured Kinetic Model of *Aspergillus niger* Growth and Substrate Uptake in a Batch Submerged Culture. *Br. Biotechnol. J.* **2014**, *4*, 970–979. [CrossRef]
39. Kara Ali, M.; Hiligsmann, S.; Outili, N.; Cherfia, R.; KacemChaouche, N. Kinetic models and parameters estimation study of biomass and ethanol production from inulin by *Pichiacaribbica* (KC977491). *Afr. J. Biotechnol.* **2017**, *16*, 124–131. [CrossRef]
40. Zajšek, K.; Goršek, A. Modelling of batch kefir fermentation kinetics for ethanol production by mixed natural microflora. *Food Bioprod. Process.* **2010**, *88*, 55–60. [CrossRef]
41. Pazouki, M.; Najafpour, G.; Hosseini, M.R. Kinetic models of cell growth, substrate utilization and bio-decolorization of distillery wastewater by *Aspergillus fumigatus* UB260. *Afr. J. Biotechnol.* **2008**, *7*, 1369–1376. [CrossRef]

MDPI

St. Alban-Anlage 66

4052 Basel

Switzerland

Tel. +41 61 683 77 34

Fax +41 61 302 89 18

www.mdpi.com

Foods Editorial Office

E-mail: foods@mdpi.com

www.mdpi.com/journal/foods